Muffin the Mule

Commemorating
60 Years of
Muffin the Mule
With Memories and Memorabilia

By Adrienne Hasler

Dedication:

It is said that we do not stop playing with toys because we grow old but, that we grow old because we stop playing with toys.
(Source unknown)

This book is in memory of Sally McNally.

© 2005 Adrienne Hasler

Illustrations drawn in the style of Neville Main by Paul Robbens. Tel: (01628) 672140

All rights reserved.

No part of this book may be reproduced, stored in a retrieval system or transmitted in any form or by any means, electronic, mechanical, photocopy, recording or otherwise without prior permission of the publisher and copyright owner.

Designed by EH Graphics
East Sussex. Telephone: (01273) 515527
Printed in the UK

Published by Jeremy Mills Publishing Limited
www.jeremymillspublishing.co.uk

Contents

Acknowledgements 4

The History of Muffin the Mule
Introduction 5
Team Muffin 9
Jan Bussell 13
Ann Hogarth 17
Annette Mills 21
Alfred Gilson 27
Sally McNally 29
Far-flung Muffins 41
More about Muffin 43
Muffin the Mule Collectors' Club 47
Collectors' Club Charities 51

Memories
Newsletters 1-14 53

Memorabilia
Memorabilia Guide 77
Memorabilia Rarity Guide 78
This Is Not Muffin 101
Modern Muffin Collectables 103
Muffin Syndicate 107
TV Comic Memorable Events 109

Photography Credits 113

Bibliography 115

Index 117

Acknowledgements

My thanks to Derek McNally for his unstinting support, additional information and photographs.

To Robin Greatrex for his advice and help in liaising with Maverick Entertainment Group plc and to Maverick Entertainment Group plc for their support in this project. Mike Gilson, for his information relating to the creation of the 'Moko Muffin Junior' by his father, also to Mike Cooke and Patrick Talbot for their help in supplying additional information about this toy and many others.

To Sebastian Wormell, Company Archivist for Harrods Ltd for scouring Harrods catalogues for me and to the staff at HAT (History of Advertising Trust, Norwich) for research into the Selfridges archives. Also Kyla Thorogood, Senior Researcher BBC Information & Archives - Commercial Services. To Stephen Smith of the Fairground Society (and member of the Muffin the Mule Collectors' Club) and to Adrienne Walsh (Club member) who provided me with information and led me to Stephen. My thanks to those collectors who lent toys to be photographed for the memorabilia section.

Photographic acknowledgement is listed separately.

Finally, to my family who have been increasing my collection whenever possible with help from Steve Penny, Mike Cooke and Mark Diamond.

All errors and omissions are mine and I apologise for these in advance!

Introduction

This book was to have been the result of collaboration between Sally McNally (nee Bussell) and myself, Adrienne Hasler. Sally was the daughter of the creators of the Hogarth Puppets - Jan Bussell and his wife, Ann Hogarth.

In the mid-1990s, a Channel 4 television series called "Collectors' Lot" invited Sally and Muffin to appear. This was my first sighting of Muffin since my childhood in the 1950s.

Sometime later, in 1999, I, too, was asked if I might like to appear in an edition of "Collectors' Lot". Themed 'the 1950s', I was to be filmed with my collection of Muffin memorabilia.

This appearance resulted in my creating the Muffin the Mule Collectors' Club with the very active support of Sally and her husband, Derek.

In 2001 I spoke to Sally about my suggestion of collaborating on a book. We could record Sally's memories of her parents' work and her own travels with Muffin and combine this with my record of the merchandising in my collection (and those items missing from my collection!).

Her enthusiasm for the job in hand meant that Sally decided to take a creative writing course in preparation for her task. Unfortunately family illness delayed the start of her work.

At the end of 2003 Sally had some great news. She had been approached by a company keen to re-market Muffin for today's children and had signed a contract that would allow this to happen. We were excited to learn that Muffin was to make his television return to the BBC in time for the celebrations of the 60th anniversary of children's television.

Sally and her husband, Derek McNally, were involved in the early stages of the new programmes. Muffin was to appear in animated form. This was an incentive to get organised and once again we met to put together our thoughts.

Sally's own sudden ill health meant that she was not able to start the project. Unbeknown to us, she was terminally ill. Fortunately the company with whom she had the contract managed to produce some of the animation for the new series and Sally saw the beginning of Muffin's second television career. She had the marketing schedule and was able to see how the venture would develop. Sadly she died at the end of May 2004.

Having spoken to her husband, Derek McNally, I took the decision to go ahead on the book by myself. I would use Sally's articles from the newsletters and combine it with information about the original creative team and create, as far as I could, a record of the merchandising produced in the 1950s.

Unfortunately papers relating to the Muffin Syndicate seem to have been lost during the Bussell house moves and so I have done my best using vintage advertising to try to date items. So although the creation of this book has been a solitary affair I have been supported and encouraged by my family, Derek McNally and not least by various Club members.

One such member has called me 'curious'. I've always assumed that they recognized my inherent curiosity (as opposed to my being strange!) and it is this curiosity that has prompted me to approach the book with enthusiasm.

Having decided to write the book I then had to think about a format. It seemed appropriate to update the historical information about Annette Mills and the Bussells. For many people Annette was Muffin's constant companion. Although a credit to the programme and her skills Ann Hogarth seldom registered with children and to this day people are often confused between Ann and Annette.

Sally's section was easy. Each newsletter had a contribution by Sally and I have kept them in the order in which they were written.

The memorabilia/reference guide has proved more problematic. Muffin merchandising was produced by a variety of companies and there is a singular lack of information on some products.

Ann and Annette set up the Muffin Syndicate after a somewhat nervous start. After all licensing and merchandising was a totally new concept in the world of television. Frustratingly for me many of the original companies no longer exist as independent companies or have no records relating to this period. The Muffin Syndicate itself fell into disuse a few years after Annette's death and Muffin's return to the travelling theatre. It seems that records no longer exist as I have not yet come across any although maybe tucked away in the Blake household (Molly Blake being Annette Mills' daughter) might be the documents I seek.

Many collectables' guides offer price guides. Some offer rarity guides and some offer both. Muffin merchandising is relatively limited compared to ceramics, furniture or the production of wares from one factory and prices are further influenced by the advent of internet auctions.

INTRODUCTION

In the 'good old days' (prior to internet auctions) most items would reach a collector via a dealer or a specialist fair. The dealers sourced their stock through the conventional auction rooms. It would be reasonable to say that whilst some private buyers attended auctions the majority of sales were to dealers. Prices were therefore achieved on the basis of commercial consideration. The dealer would need to factor into their bid their expenses and possible profit. With the advent of the internet auction sites and a current preoccupation by television companies with collecting and auction rooms this has changed. Prices are affected by the fact that collectors bid with their hearts not their heads!

Some bidders and buyers do find bargains but some of us, like me, can be tempted to lose their normal commonsense and pay large sums for some of the relatively common items. Prices are volatile and will probably be outdated even before this guide is published so I decided that the collectables section should be based on rarity. Rarity itself is subjective. Without the knowledge of actual production numbers one can only make an educated guess. My 'guesstimates' are made on the basis of how many of the items I have come across since starting my collection.

No doubt some of you will disagree with me. In that case I'll be updating the guide next year.

For now I hope that you will enjoy the journey chronicling the adventures of Muffin the Mule.

There are not many television personalities whose work has taken them from Europe to Africa, from Australasia to the Arctic Circle and whose career spans some seven decades.

It has been suggested that Muffin the Mule might not fit into today's world. However, in October 2004, Muffin and Annette were credited as the inspiration for the promotional video created for the new single released by an up-and-coming artist called Dizzee Rascal! His video director, Dougal Wilson, writes for the web page: "I've always been interested by old children's TV shows. My main source of inspiration for this video, however, was "Muffin the Mule" which starred Annette Mills, who had a beautiful 1950s BBC demeanour. The show has a charming innocence about it. I thought it could be amusing to contrast this with the story of Dizzee's origins in Bow and Hackney."

Not only did the show have a charming innocence it also featured many different types of animals. Muffin's world was as diverse as those of children today and I for one am sure that he will be as loved by a new generation of children as he was by my generation.

Muffin the Mule

Team Muffin

Photo - courtesy Derek McNally

So, if you are sitting comfortably, I'll begin with a little background information and then a closer look at the individuals who were the major human influences upon the career of Muffin the Mule. I can do no better than to introduce the foremost members of 'Team Muffin' in much the same way as one would provide a cast list and résumé.

However, in selecting those who were key elements in Muffin's international success, I am neglecting Stanley Maile, Jack Whitehead and Pauline Jackson, all of whom contributed to the 'backstage' successes of the Hogarth Puppets, including Muffin.

Both Jack Whitehead and Stanley Maile already had established reputations as puppet makers/carvers. Having joined the Hogarth Puppet Theatre they employed not only these talents but, according to Jan Bussell in 'Puppets Progress', became proficient in 'stage management and scenery moving'. Stanley Maile also showed a skill as an illustrator. He provided the drawings for Ann Hogarth's The Red/Blue/Green and Purple Muffin books.

Pauline Jackson worked as assistant to Ann Hogarth on virtually all of the Muffin programmes for the BBC and also toured with the company, other assistants being called upon as required for special productions.

One further omission is Neville Main. His works include his own Muffin stories, both as illustrator and author. He illustrated the 'Merry Muffin' series of books and then created and drew the puppetry for the first Muffin story ever to be animated. This was the story 'Muffin and the Reluctant Carrot' which employed the new form of animation invented by Jan Bussell.

I would like to acknowledge the talents of the puppet creators, the carvers, the stage managers and the other puppeteers and apprentices who worked with and supported the enviable reputation for high quality performance by the

Hogarth Puppet Theatre. There are professional puppeteers and writers (including Jan Bussell and Ann Hogarth) who have already written in detail about these people and I can highly recommend their books to readers.

Jan Bussell records the contributions and recollections of those who were actively involved at the time including, in a brief mention in 'Through Wooden Eyes' (see later), Jane Phillips.

I mention Jane for two reasons. Firstly, Jane was apprenticed to Ann Hogarth as a young woman and went on to create her own highly successful puppet company as well as performing as puppeteer on several notable television productions. After Ann Hogarth's death Jane was charged with the care and placement of many of the Hogarth puppets. For the past few years she has been responsible for producing the exhibitions of Hogarth Puppets around the country.

Secondly, Jane is a Muffin the Mule Collectors' Club member and has been a great source of information to me, lending me her precious Muffin related scrapbooks and was responsible for compiling the entry on Jan Bussell and Ann Hogarth for the 2004 edition of The Oxford Dictionary of National Biography (OUP2004).

Knowing that Sally's ambition was to bring together puppets that had been spread far and wide over the past 40 years, Jane also managed to track down 'lost' puppets. When I saw Sally for the last time she gave me Otto the Octopus to hold… Jane had just sent him to her.

The following pages will introduce you more fully to the people involved in the success of Muffin the Mule, but for now a brief description.

The aforementioned Jan Bussell, Ann Hogarth and their daughter, Sally McNally, made up the Bussell family. Jan Bussell was both involved in puppet theatre and the early days of television. He worked with John Logie Baird and became a BBC television producer. Ann Hogarth, his wife, trained at RADA as an actress but found her forte was as a puppet manipulator. She was highly respected by her peers. Ironically, given the success to come, puppetry was still very much a medium targeting adults and the combined talents of the Bussell family created the Hogarth Puppet Theatre.

Sally, their daughter, became a manipulator with skill rivalling that of her mother. Her early schooling provided equal talents in dance, stagecraft and management.

Annette Mills was another multi-talented woman whose highly successful career had been blighted by massive injuries sustained in a car accident during the blackout in World War II.

Her new television career brought her even more success than her previous one!

"Team Muffin".

As I have already mentioned, the Bussell family and Annette Mills were the visible members of the team. In addition to the above members I have included Alfred Gilson. Not a name with which most readers will be familiar. Despite this, Alfred was an important component in Muffin's success as a television icon! Alfred designed, developed and initially manufactured the Moko Muffin Junior puppet that sold in its 1000's. I feel it appropriate that he gets a mention.

Muffin the Mule

Jan Bussell
1909 – 1985

Photo: Jan & Kirri the Kiwi - courtesy of Derek McNally

Although arguably the most famous characters, Muffin the Mule & his fellow Hogarth puppets, they were only part of the extraordinary repertoire of puppets created by the combined talents of Jan Bussell and his wife, Ann Hogarth. In a way Jan Bussell's choice of career seems an unusual one. His father had been a clergyman (although he was killed during WW1) and one of his forebears, the Rev J. Bussell, is commemorated in the great west window of Newark's medieval church. The window commemorates the fact that when the vicar of this church he was responsible for the major 19th Century restoration of the building. He became Canon of Lincoln in 1835.

A few years earlier other Bussell family members ventured further afield. The town of Busselton in Western Australia is named after the part of the family which arrived in Western Australia in 1830. In the early 1950s, a visit by The Hogarth Puppets to Australia apparently passed by without the distant Bussells meeting. However, in 'The Puppet Master' (June 1956), the monthly meeting is reported to have heard of the latest Australia/New Zealand trip during which Jan Bussell met 200 cousins!

Jan Bussell's interest in puppetry was, in his words, "a continuation of a childhood toy theatre which I have endeavoured to turn into an art". When one looks at the range of puppets and productions created by the Hogarth Puppet 'troupe' it is quite clear that the professional careers of both Jan Bussell and Ann Hogarth were pursued with that goal firmly in mind. And so it was that in the 1920s, the London Puppet Theatre gained a new apprentice and then, in the early 1930s, whilst working with the London Marionette Company, Jan became involved with the experimental work of John Logie Baird which included producing television programmes using puppets. There were great advantages in using puppets in television (they were small, had their own sets and theatres etc) and consequently the puppet companies were highly sought after.

Jan Bussell managed to combine his passion for the puppet theatre with the new medium of television and he worked for the BBC from its infancy, participating in the puppet theatre/television collaboration.

Having been instrumental in securing the booking with John Logie Baird he was credited with being responsible for the first television broadcast using puppets in the world. In 1932 he and Ann Hogarth set up the Hogarth Puppet Theatre and by 1936 both he and Ann Hogarth were set on their life's work planning and producing puppet plays. These included opera and shows for both adult and child audiences.

Sophisticated productions, designed for adult audiences, included 'Master Peter's Puppet Show'- an opera based on Don Quixote; the beautiful shadow puppet production of Oscar Wilde's The Happy Prince, and the orchestra and circus creations, to ballet scenes and portions of Macbeth. The wit and humour of both Bussells can be illustrated by the gloriously named Anti Cyclone and Deep Depression - both puppets from The Flower Ballet.

The company travelled extensively, touring Czechoslovakia, New Zealand, Australia and South Africa. Later tours included the USA and Canada with what was probably the most northerly production staged within the Arctic Circle.

The diversity of countries was matched by the range of puppets employed - from the stringed puppets, i.e. marionettes (of which Muffin was the most famous); shadow puppets - most memorably - the beautiful creations by Lotte Reiniger for Oscar Wilde's 'The Happy Prince' and the glove puppets used for the television version of Alison Uttley's 'Little Grey Rabbit', to rod puppets worked in the late 1960s.

Perhaps one of the most striking and extraordinary puppets is the 'Picasso' puppet. This puppet has a wonderful 'Picasso' style head of rubber on a body which is, in part, worn like a long evening glove.

Photo: Picasso puppet from Meister Des Puppenspiels

The puppet was created for the adult performances with an accompanying verse which, presumably, was to counter the notion, based on Muffin's success - that all Hogarth Puppet Theatre performances were for children.

The following verse (only partially reproduced) was written by Ann Hogarth to be spoken when working the Picasso puppet:

> *I am a Puppet of a Portrait of a Poet*
> *Painted by Picasso*
> *Presented with a Purpose*
> *The Problem of Puppets at the Present Period*
> *Is the Prejudice of the Populace we are Puerile.*
> *This is Patently Piffle For the Proper Performance of the Puppet*
> *Is to Portray the Peculiarities of People - Prince, Proletarian, Priest or Politician.*
> *And for these Portrayals to be Properly Appreciated*
> *Presumes a Pre-knowledge by our Public of the Persons so Presented*
> *This the Precious Poppets of Proud Parents cannot Possibly Possess...*

I was privileged to be given a performance of the puppet (and the verse) by Sally McNally. Although the puppet is an extraordinary creation it is quite compelling and most elegant.

Following the incorporation of Hogarth puppets into the Sunday evening children's programme with Annette Mills, the Bussells combined their work for the BBC with their ongoing tours. For a fascinating first hand account of this time, you would do no better than to read Jan Bussell's books 'Puppet's Progress' and 'Through Wooden Eyes'. His books illustrate both the passion and the professionalism consistently demonstrated by both Jan Bussell and his wife. When one reads of the somewhat less than luxurious conditions in which they lived and worked it is hard to see the attraction of owning a travelling theatre group!

Although the main medium of their work was live theatre and television, the Hogarth Puppets also took to film. In 'Puppet's Progress' Jan describes the approach by Parthian Films in 1949. The decision to make films for American television is one for which the Muffin fans of yesteryear must be grateful. We owe the existence of the majority of surviving Muffin stories to Parthian. Neither the BBC, nor the later ATV, recorded the programmes at the time of transmission. It is known that some twenty-six Muffin films remain in existence and some copies are in private collections.

Jan Bussell had an illustrious career. He combined skills as a behind-the-scenes radio and television producer with his role as a showman. He could perform with Muffin, hold the attention of fellow puppeteers whilst lecturing and still find both time and enthusiasm to work on behalf of professional puppet organizations. He was the editor of The Puppet Master, the journal of the British Puppet and Model Theatre Guild between 1956 and 1960. He also held the presidency of UNIMA (the Union Internationalle de la Marionette) for two terms (8 years).

Alongside such activities Jan Bussell found the time and creativity to invent a new form of animation.

The British Puppet and Model Theatre Guild newsletter dated September 1956 records, under 'Items of Interest', that Jan Bussell writes that the Hogarth Puppets have now completed their 26 Muffin programmes on ATV and also their open air performances in 45 London parks. He goes on to say that he has discovered a new method of making colour cartoon films with invisible live animation and plans to spend a good deal of time on this during this autumn.

He demonstrated this technique in a short Muffin film called 'Muffin and The Reluctant Carrot'. The story was scripted by Neville Main and he also made the puppets for the film. This film has been shown to members of the Muffin the Mule Collectors' Club and received an enthusiastic response. It seems fitting that, twenty years after Jan Bussell's death, Muffin is set to appear, once again, in animation.

Ann Hogarth
1910 – 1993

Photo: Ann & Peregrine - courtesy of Derek McNally

Ann Hogarth originally trained as an actress studying with the prestigious Royal Academy of Dramatic Arts (RADA) but considered herself 'a very bad actress'. She moved into stage management and it was whilst working in this capacity that she first met and subsequently married Jan Bussell. Although puppets had not been a part of her early life, it was through Jan Bussell that she discovered her natural ability – as a very fine puppet manipulator. Indeed, she became a 'hidden' actress using her puppets to do the acting for her. She was able to create a personality through manipulation in a way she felt unable to achieve in front of an audience. This skill is much in evidence when one sees how Muffin or, as he was formerly known, 'the Kicking Mule' behaves on stage. As one half of the circus act the Mule is bad tempered and not at all like the naughty school boy that is Muffin. The same puppet with two distinct personalities.

In his book 'Puppet's Progress' Jan Bussell notes that "most people regard her as the greatest English manipulator". It is certainly true that her erstwhile 'bad' acting skills blossomed in her role as puppet manipulator. With their combined talents for production, direction and stage management the Hogarth Puppets were to become both extremely successful and highly respected for the academic pursuit of covering worldwide heritage of puppetry. Both Ann Hogarth and Jan Bussell wrote books for both professional and amateur puppeteers. Ann herself wrote several Muffin books and a book on puppetry for younger children. It is suggested that one of Ann's proudest moments was to see copies of her Muffin books translated into Russian and Latvian.

Ann designed numerous puppets including the previously mentioned wonderful 'Deep Depression and Anti Cyclone'. She knew exactly the personality of each creation. It seems that she also had a clear idea as to the personalities of others. It is recorded by Sally McNally (her daughter) in her tribute to Ann Hogarth (The Times 24th April 1993) that '..she gained the reputation in the Puppet Guild of being the rudest woman in England'! Despite this somewhat dubious reputation Ann exhibited not only a talent for manipulation but also for writing witty scripts and verse. It is often forgotten that it was Ann who wrote the scripts for the fortnightly Muffin programmes. Annette Mills would then compose the songs to meet the requirements of the stories and her own words. Although Ann considered herself a relative 'latecomer' to puppets, her enthusiasm for puppetry continued well into her eighties. I suppose it must be said that, although most strongly identified with her little Mule, like other actors who become so identified, she could express her frustration in that her work with other incredibly sophisticated types of puppets and the resultant performances was often rather overlooked!

Her collaboration with Annette Mills spanned eight years. This was a highly successful combination. The two women would meet after Ann had produced the scripts and Annette would write the songs for the live programmes transmitted on alternate Sundays. Together they were instrumental in producing the first international star of television...Muffin!

Their combined creativity inspired stories that were visually and musically enchanting and became compulsory viewing for adults and children alike. It was said that Annette appealed to the fathers and Muffin to the children!

"Now then, Muffin!" The mule in a tight corner between Annette and Ann Hogarth

Photo: Ann & Annette from Bandwagon

Advertising the Hogarth Puppets' forthcoming appearance at the Theatre Royal, Bath, the programme states "A witty and light-hearted entertainment for all". The Daily Mail, reporting on a previous performance, is quoted (as stated in the programme for the Lyric Theatre, Hammersmith) as noting "This is the sort of show to which really intelligent parents will take their children. We might also say that it is the sort of show to which really intelligent children will take their parents".

Alongside the fortnightly television performances and the touring shows were the revues and pantomimes. Muffin appeared alongside a young Audrey Hepburn in Cecil Landau's Christmas Party (the resulting theatre programme is now highly sought after by Hepburn collectors) and the Queen of Hearts.

Picture of Annette & tarts - courtesy of Sally McNally.   Poster from the play, from TV Weekly

What also has to be remembered is that Ann Hogarth was required to perform with Muffin for considerable periods of time and Muffin is not a lightweight mule!

It is true that Muffin had originally just been known as the 'Kicking Mule', but Annette's inspiration in naming him inspired a new personality for Ann to develop.

Behind Annette's sophistication was the remains of a lonely and shy child...this gave her an understanding as to how children related to Muffin. Annette's Muffin the Mule records were proving to be a great hit in America and there was a suggestion that Muffin might cross the Atlantic (with the 'Parthian' films having originally been made for America, this would seem a strong possibility). Muffin's success on stage and screen meant that merchandising deals were forthcoming. During a television interview at the Pebble Mill studios in 1985, Ann Hogarth recalled the licensing agreements offered to her during the early Muffin years. She noted that, originally, the manufacturers had informed her that they would expect her to pay them to have Muffin products manufactured and sold. After some discussion with Annette Mills, whilst sitting in Lincoln's Inn Fields, it was determined that was not the way things should work and eventually the merchandising was protected by the setting up of The Muffin

Syndicate in 1949. The Syndicate licensed many products and information about these appear in the memorabilia guide section. Sadly no records have been found relating to all of the possible licences. According to Ann Hogarth, during the interview for the programme, to make money from merchandising agreements one needed about 100 commodities. She commented that 'they were getting there' when Annette Mills sadly died.

Annette Mills, Ann Hogarth and Jan Bussell at a book-signing session. Photo: courtesy of Derek McNally

Ann and Muffin were still in demand when, in 1990, the BBC closed Lime Grove Studios. It seems that The Lime Grove Studios had opened on 21st May 1950 with a meeting between Mrs Clement Attlee and Muffin and so the producers of 'The Late Show', when making 'The Lime Grove Story' in 1980, decided that Muffin should be there as the doors closed for the last time. Thus Muffin was invited to take part in 'The Lime Grove Story'....more of that later when Sally recalls that particular day.

I had assumed that Zebbie the Zebra had been created for tours of South Africa but I discovered that I was wrong! In an interview for 'Illustrated' magazine (December 1953), Jan Bussell explains that Zebbie was a joke in a circus act. A puppet called Colonel Poona, riding an elephant, would shout 'Look, there's a zebra crossing!'

Well, I'm sure it sounded funnier then!

It must be said that the readers of the Illustrated London News (12th Jan 1952) were treated to diagrams instructing them 'How to use a Zebra Crossing'....

Ann Hogarth and Muffin returning from South Africa. Photo: courtesy of Derek McNally

Annette Mills
1895 – 1955

Photo: Annette & Muffin - courtesy of Derek McNally

Annette Mills, older sister of Sir John Mills, trained as a concert pianist and then changed career becoming an exhibition dancer. She partnered Robert Sielle, whom she subsequently married. In John Bull (9th September 1950) it is reported that 'after the Astaires, she and her partner ranked as one of the highest paid pairs in the world'. Her dancing career saw her introduce the Charleston to Britain after seeing it performed in New York and their partnership took them around the world including a fateful trip to South Africa. Whilst on tour in South Africa Annette broke her leg. Unable to dance, she returned to her piano and composed songs for revue and cabaret. Amongst her compositions was 'Boomps -a- Daisy' which featured in the highly successful 1938 Broadway show "Hellzapoppin". This show ran for 1,404 performances and was one of Broadway's longest running musicals of that time. It appears that it was also an anti - Hitler song which apparently got her onto a Nazi 'hit list'. Another was 'I'm An Old Norman Castle (1940) which describes the happenings to a building taken over by the military during the war.

With the advent of war Annette Mills did her part as an entertainer. It was the result of a road accident, in 1942, whilst returning home from entertaining troops at North Weald air base (her car was in collision with a military vehicle in the blackout) that events would lead to her becoming, in her own words, 'a stooge to a Mule' (TV Mirror 29th August 1953). Importantly, her resulting injuries were to keep her hospitalized for over two years (1943/46). Annette's injuries were considerable. The accident caused long-lasting damage to her hands and legs and this would eventually have a great impact upon her and influence her subsequent involvement in the formation of the Muffin Club.

Once recovered (although she remained in pain for the rest of her life), Annette turned to children's television and in 1946, shortly after television resumed transmission after the war, she was appearing fairly regularly in a programme called, rather unimaginatively, 'For the children'……… seated at a grand piano. Her Sunday evening performances were live and she decided that her piano could be utilized by becoming a stage for puppets.

In TV Mirror 29th August 1953, containing the epithet 'Stooge to a Mule', Annette describes how she came to arrive at the home of the Bussells. Annette had the image of small creatures walking across her piano top and auditioned the kicking mule and his clown partner for this job.

(Sally recalls this meeting in her newsletter reminiscences).

Photo: Kicking Mule. Courtesy Derek McNally

Annette gave each of the puppets a name and they took their turn on Annette's piano. Sadly, for Crumpet the Clown, the response to his performance was not good. Muffin, on the other hand (or, should that be, hoof), became an overnight success.

Annette was in no doubt that it was the unique and inspired partnership between her and Ann Hogarth that produced the magic that was the Muffin programmes. A well-honed routine ensured that between October 1946 and December 1954, with a few exceptions, for example, when Muffin & family took off on their travels around the world, Annette (and Ann!) appeared with Muffin and his friends on alternate Sundays. Ann Hogarth would create the story and send it to Annette. She in turn would write the music and lyrics and prepare her dialogue. Watching the programmes one is struck by the calm way she "ad lib's" if something untoward happens with a puppet. In one scene Peregrine the Penguin's foot becomes tangled and she calmly untangles him whilst speaking to him in a matter of fact way as one would a child.

In this article Annette describes her ideas as to the success of the little mule. She wrote: "Perhaps because he is, to me, any small boy. His feet are planted firmly in the nursery or in the kitchen, but his soul is free to soar into realms of fantasy. He is loving, bumptious, busy, lazy, greedy and sentimental. He is utterly in earnest. I love him dearly but I have no illusions about him. In the words of my brother John, I am now merely a 'stooge to a mule'. "

Stooge she may have felt but Muffin's success brought Annette new challenges and popularity. In 1949 she won the Television Society's silver medal and was voted outstanding personality/best television artist.

She wrote Muffin stories including the six books in the Muffin the Mule range, illustrated by her daughter, Molly Blake. Annette also appeared with Ann and Muffin at stores, civic events and, with the inception of the Muffin Club, in hospitals and children's homes as the residents took possession of the televisions and radios bought through money raised by the club.

In addition to Annette's appearances with Muffin she created her own hand puppets. According to 'Illustrated' magazine (10th Dec 1955) Prudence was born on 7th June 1950. Prudence was an elegant cat. She had a sister Primrose and various other companions. Eklan Allan recorded, in John Bull (9th September 1950) that Prudence had a wardrobe of nine dresses including two that were copies of Annette's Paris models. Prudence played a virginal (an early and rather elegant keyboard instrument played in a similar way to a piano but related to the harpsichord). Despite the constant pain from her hand injuries Annette was able to allow Prudence to remove a tiny hanky from her pocket, use it and then replace it. It is sometimes suggested that Prudence appeared with Muffin. This is not so. Prudence and her companions had their own programme. Muffin, of course, was a Hogarth puppet and Prudence was a Mills puppet.

Huge crowds would gather at every event and Annette and Muffin's autographs were highly sought after. If one managed to get Ann, Annette and Muffin's autograph, that was THE prize.

The love and affection felt for Annette was graphically demonstrated following her death in 1955. There were huge numbers of mourners at her funeral and many floral tributes— including those created as Muffin.

Annette Mills had died in early 1955 and, ironically, because of her huge popularity and identification with Muffin, the BBC took the decision not to continue with the programmes. People remembering her death speak of the sadness they felt and the sense of loss. Not only losing Annette Mills but also of losing their programme and of course, Muffin. Indeed, an article in a newspaper from 1955 bemoans the decision of the BBC not to continue with Muffin after Annette's death. Ann Hogarth is reported to have suggested teaming up with Molly Blake but the BBC was adamant. The writer concluded that the children would have understood perfectly if Muffin appeared without Annette but now they were losing two friends because of the attitude of the BBC.

It seems that Ann Hogarth was right and the BBC wrong. The Muffin stories have been remembered by those whose childhoods started in the decade following the end of World War II and are still remembered today almost 60 years later with great affection . Following her mother's death Molly Blake took over writing Muffin stories for later annuals and in the autumn of 1955 she brought Prudence Kitten back to television. Prudence appeared in a version of Sleeping Beauty televised on 2nd January 1956 and in subsequent programmes.

Muffin moved to ATV for 26 shows and gave him what was described as 'a new personality and new setting'. Initially, Jan Bussell took the role as Muffin's human companion, after all he had often appeared with Muffin whilst on tour without Annette or as her substitute. However, after two episodes, his daughter Sally took on the role. The contract ended at just the right time for Sally - her Saturday commitment to Muffin being overtaken by her wedding to Captain Derek McNally on 1st September 1956.

Muffin then 'retired' from full time television although, of course, he continued to make guest appearances and continued his international trips with his fellow Hogarth puppets.

It was at this time that Jan Bussell reported to The British Puppet and Model Theatre Guild that 'he had discovered a new method of making colour cartoon films with invisible live animation and planned to spend a good deal of time on this during this autumn'. As I have already mentioned he subsequently produced 'Muffin and The Reluctant Carrot' a story created and modelled by Neville Main. The animated Muffin (and his friends) showed all of the characteristics that Annette had identified some three years earlier.

Fortunately many children were able to see Muffin in his role as a 'star' in the Hogarth Puppet Theatre and during the 1960s and 1970s Muffin and his fellow Hogarth puppets continued touring the world from Eastern Europe to Australasia and the continent of North America.

NZ High Commissioner Mr FW Doidge & Kirri the Kiwi (NB note says KIKI!) 1953. Photo: courtesy Derek McNally

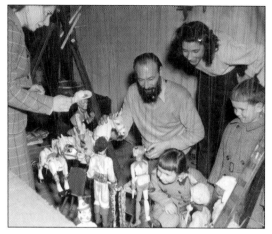 

Children in the UK and South Africa getting acquainted with Muffin. Photos: courtesy Derek McNally

In September 1949, in the publication *Bandwagon*, Sylvia Duncan records the impact of the little Mule who started life as part of a circus act. She notes the numerous letters which arrived addressed to 'The Potting Shed, Bottom of Annette's Garden, London' and the gifts that arrive for him. She commented that the BBC is scared to do the obvious and run a 'paint Muffin' competition. Even in 1949 whilst a great proportion of the UK was without television Muffin's fame had already spread to the US. Records of Annette Mills' songs had been so successful that films were being made for American television including two cinema films.

The Muffin Song Book had already made an appearance and toys were scheduled for Christmas 1950.

Sylvia Duncan wrote... 'It was certainly nobody's guess that the first person to ever win an international reputation from television would simply be - a little wooden mule'.

Robert Dane-The Sketch (9th November 1949) wrote:

"The amazing thing is that each of these creatures - right down to Doris, a quiet homely little fieldmouse if ever there was one, has been invested with a personality which is as clear cut and as certain as that of many live TV performers, and a great deal more interesting than most...But for us hard-boiled old cynics that we are Muffin's Sunday appearances in Children's Hour are alone well worth the price of a set."

Such was Muffin's success that toy manufacturers and book publishers became eager to create toys, books and household items that would appeal to children and parents alike. In 1949, The Muffin Syndicate was formed to permit manufacturers to apply to license products relating to Muffin and his friends.

In the last section of this book is a photographic record listing many of those items. One of these toys was the Moko Muffin Junior, a die-cast toy. He was created and manufactured but not by a large international or even national toy company. Instead, he arrived via a company that started in a garage workshop!

Alfred Gilson
1912 – 1980

Photograph of Alfred Gilson and Annette Mills' invitation to Selfridges letter.
- Courtesy Mike Gilson

Alfred Gilson, the creator of the Moko Muffin Junior, began his toy manufacturing business toward the end of WWII. In fact Alfred became a die-cast toy manufacturer by accident! He was, by training, an electrician and during the war was employed in this capacity as an electrician working on Liberty ships. For those of us who did not know (probably most of us) 'Liberty ships' were civilian built merchant ships built during World War II. A vital component of Britain's naval resources, they carried cargoes of grain and mail, ore and ammunition, trucks and troops in the huge convoys that crossed the Atlantic between the US and Britain.

Alfred was also a part-time ambulance driver during his off-duty hours. Living with his young family in North London, Alfred's toymaking began as a result of the shortage of everything, especially frivolous items such as toys. His sister-in-law purchased the mould for her to make gifts for the children in the family and he saw bigger opportunities and bought them from her. Having purchased the mould (plus a small cast-iron pot and a ladle) from her he went on to make more soldiers.

It seems that his toys were shown to other children and their parents requested similar figures. Given the austerity of post-war Britain it was the perfect time to produce toys. People were hungry for items such as this. Although the procedure was very crude (I believe that the family garage was the site of the first manufacturing facility) and the resulting figures were fairly primitive they were, nonetheless, much in demand items by people who wanted toys for their children. Little did anyone realize where this small endeavour would lead. His son remarks that, even at this limited level, there was considerable demand because he hired a worker to pour these toys for him while he went to work!

Alfred Gilson then created a legend…a die-cast version of Muffin the Mule. It is not clear quite how this happened, but it did, and in July 1950 Annette Mills wrote to Alfred Gilson to invite him to join her at the promotion of the Moko Junior Muffin puppet at the Selfridges store in London. Unfortunately, he did not receive the letter in time and missed the occasion, but his newly created company, A. Gilson Ltd., manufactured the puppet until Alfred sold up and left the UK. The factory site in Dalston, East London was large and so Alfred rented out a portion of the building to another toymaking company. This company had been set up by Rodney Smith and Leslie Smith and was called Lesney Products & Co. Later they would be joined by Jack Odell and create the wonderful Matchbox toys. Alfred's son, Michael, recalls the early success of the Moko Muffin. He writes 'my father drove a Razor Edge model Triumph. I still remember the licence plate was JAR402. It was for this car that he had a Muffin head chromed and mounted on the top of the radiator cap'. The puppet was marketed by Richard Kohnstam under the name J. Kohnstam Ltd. Kohnstam's company trade name was 'Moko' and all of the distinctive die-cast Muffin puppets were named "Moko Muffin Junior".

The Gilson puppets were only produced for a very short period - in 1950 - before the British Government was forced to ban the use of zinc for toys - as a result of the onset of the Korean War. Frustrated by this, Alfred cast his eyes on new horizons; he emigrated with his family to Canada but eventually settled in California. Having arrived in the US Alfred went back to his original trade as an electrician and continued as such for the remainder of his life. Meanwhile, back in the UK, supplies of zinc gradually became available once again and from mid-1951 Lesney Products Limited took over the production of Muffin Junior for Kohnstam.

Lesney made the puppets until 1957. It is estimated that some 60,000 puppets were sold. Although neither the Gilson nor Lesney name appears on the puppet box, the Lesney name is that which is more generally associated with the puppet and for Lesney toy collectors a pristine boxed Moko Muffin Junior is a desirable acquisition. For some reason, the boxes are marked as by permission of 'the Muffin Syndicate Ltd. whilst other items are marked 'The Muffin Syndicate'.

In 1949 Selfridges were advertising a wooden Muffin the Mule. This item seems not to have been a licensed product and its manufacturer remains unknown...for now. It seems that the Moko took over from the wooden version which is now extremely rare.

Wooden Muffin from The Sunday Graphic, November 27th 1949. Muffin cost 19/- (95p).

Sally McNally
1936 – 2004

Sally repainting Muffin - courtesy Derek McNally

I never had the opportunity to meet Jan, Ann, Annette or Alfred but I count myself fortunate indeed to have met Sally McNally.

Having a personal knowledge, however limited, of someone alters the way one thinks or writes about them and so this chapter differs in layout from those proceeding. I hope that this does not affect your enjoyment in reading about an extraordinary lady and I hope that my personal input does not detract from the story of Sally.

Before I launch into telling you about Sally, I'd like to tell you a little story….

Shortly after I started the Club, Sally and Derek invited Ron, my husband, and I to visit them if we were in Devon.

As you can imagine, I leapt at the opportunity and the visit was duly arranged with me clutching camera and vast quantities of film.

Sally and Derek were extremely hospitable and we enjoyed morning coffee with carrot cake and then a glass of sherry (offered in a Muffin decorated glass) before a lovely lunch. It was then suggested that we take a tour of neighbouring sites of interest. This included buildings relating to Sir Francis Drake and Ann Hogarth! Ann had spent her last years in Budleigh Salterton. Having parked their car near her mother's former flat, Sally directed us towards the High Street. As we walked she glanced about her and told me that she often felt like a school girl when visiting the town. The urge to flick imaginary pigtails and skip along the street was strong!

Looking around at the local population, I understood her comments. The population seemed predominantly elderly being early in the year – the visiting families, anticipating bucket and spades plus lots of ice cream, were yet to appear. This observation reflected Sally. She had a joy and enthusiasm for life and despite the numerous times that she had been greeted by "We Want Muffin" she even retained her enthusiasm for working her little Mule and his friends.

As might be anticipated from reading about her parents, Sally had a somewhat unconventional childhood. In her entry for the Oxford Dictionary of National Biography Jane Phillips records that the Bussells had spent their honeymoon camping and touring with the Hogarth puppets so it would seem unlikely that the next momentous event in their lives – the arrival of a daughter – would cause more than a ripple in their activities. Indeed, in her address at Sally's cremation, Alison Orchard (of the British Humanist Society) told those gathered that Sally gave her first radio interview at the age of five.

It was at this age, too, that Sally was enrolled in the Cone Ripman School (now Arts Educational School) in London. The founders of the Arts Educational Schools, Grace Cone and Olive Ripman, believed that a talent in the performing arts had to be nurtured alongside rigorous training. In place of gym and games, the pupils took ballet and acting classes. It seems that amongst other pupils who may have attended the school at about the same time as Sally are actresses Amanda Barrie and Julie Andrews, and ballerina Antoinette Sibley. Certainly the school honed Sally's numerous talents. Although, in general, I found that she was extremely reluctant to talk about herself she did tell me about her dancing at Hampton Court. She, being part of a troupe of dancers, gave exhibitions of the court dances of Tudor times to visitors.

When I asked Derek about Sally's time at school he told me that Sally missed the opportunity to appear in the 1954 film 'The Belles of St. Trinian's' as she was on tour with her parents in South Africa. Although her schooling was interrupted more than once by trips with her parents Sally made long-lasting friendships whilst at school and these friendships endured until her death. An indication of the esteem in which she was held by friends from these days was seen in the willingness of some to travel across oceans to be at her cremation. Although her father records in Puppet's Progress '- we have not taken her with the show a great deal -', Sally did make a great many public and behind-the-scenes appearances. As already mentioned, it seems that her first radio interview was at the age of five talking about being bombed by the Germans.

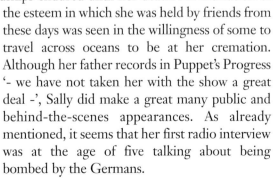

Sally, from BBC Year Book - courtesy Copyright © BBC

In the BBC Year Book 1946, there is a

photograph of Sally (reprinted here) on Radio Roundabout. She is telling the children of India about her father's puppet theatre.

By 1950 Sally was becoming more and more involved with her parents' puppet engagements. She joined her parents at Woolacombe Beach in North Devon during their month-long booking. It seems that the Bussells took advantage of the popularity of Muffin toys by selling brooches and Pelham puppets after the performances. I was able briefly to reunite Sally with one of these brooches. As you can see from the photograph, it would seem that Woolacombe Beach provided a glorious holiday as well as hard work!

See Newsletter 14th Feb 2004 for Sally's account. The plastic brooch appears in the memorabilia section.

When the Bussells created the puppets for the BBC television production of Alison Uttley's Little Grey Rabbit and their own 'The Bookworms' series Sally provided her skills as a manipulator. In July 1951 she was acting as a substitute manipulator alongside Pauline Jackson because the Hogarth Puppet booking at the Riverside Theatre, Battersea coincided with Muffin (and Ann's) television appearance.

Sally at Woolacombe Beach.
Courtesy Derek McNally

As Sally grew older, she joined her parents on trips overseas. In 1951, she travelled to Australia for the first time and she writes of this trip in her article for Newsletter 5. Jan Bussell records his own vivid memories in Puppet's Progress and so I will allow the reader to learn of this adventure directly from these sources.

1954 saw the Hogarth Puppets (and Sally) journey to South Africa. Again, this trip is wonderfully documented by Jan Bussell in 'Through Wooden Eyes' and the Pelpups (Pelham Puppets) magazine.

Suffice to say that Sally and her father appear to have enjoyed the majestic scenery (and heights) more than Ann did!

In her newsletter articles Sally told us of other occasions which stood out in her memory, for example, climbing on board an elephant in place of her mother...As you have probably begun to realize, Sally was a lively and courageous woman. These attributes brought her to the attention of the young, and very handsome, Army Captain who would become her husband.

It is told that both were attending a party given by an Irish lord. It was 1956

Sally & parents on boat - courtesy Derek McNally

and Sally was now a young woman of 20 years. This 'gentleman' made some unpleasant remarks about the cabaret and Sally berated him. It seems that his response to her rebuke was to swear. In 1956 this was considered a most offensive act in front of a woman and Sally emptied a glass of beer over him. According to Derek McNally, the glass of beer belonged to him and he promptly suggested that Sally buy him another. It is obvious that Captain McNally recognized courage and a joy for life because, just six months later, on 1st September 1956, Captain Derek McNally made Sally Bussell his wife and they enjoyed a long and happy marriage. Sally enjoyed motherhood. She and Derek had a daughter Lucinda, followed by a son, William, and she spent their childhood as a full time mother. In his address at her cremation, William spoke of Sally's enjoyment of life. He recalled jumping into puddles (Sally, not him!), family parties, including playing croquet by the light of numerous cigarettes, and her energy and expression when reading stories to himself and Lucinda. Sally and her family moved several times as Derek's work or family circumstances dictated. Eventually, in 1987, Derek retired and he and Sally moved to Devon. Jan Bussell had died in 1985 leaving Ann Hogarth alone at their home The White Barn in Whimple, Devon. Ann eventually moved to Budleigh Salterton to be nearer Sally and, as we read in Sally's recollections, Ann and Muffin were once again in demand….and so was Sally!

During the years that her family grew up and moved away Sally ran her own puppet theatre, gave numerous talks to a variety of groups and societies and worked on behalf of various organizations and charities. She was often asked to appear on radio programmes and as luck would have it (for me) appeared on television.

As Sally recalled in her newsletter articles she was frequently called upon to take on the more arduous tasks for her mother. Sally accompanied her mother to Lime Grove for the programme covering the last transmission in June 1991. Muffin's appearance at Lime Grove was quite appropriate as he had attended the opening of the studios in May 1951. However, Sally was often amused and a little frustrated at the many instances when Muffin would appear or be referred to completely anachronistically.

I was involved on one such occasion. I received a phone call from someone purporting to be from the BBC. I was told that this person was working on a 'Doctor Who' night and they wondered if they might borrow 'my Muffin the Mule'. Maintaining my politeness (despite the fact that I thought that it was someone teasing me) I explained that my Muffin the Mule was a 6" high metal puppet and that the person they really needed to contact was Sally. I rang her and explained the call and promptly forgot the incident.

Sometime later I happened to notice that there was indeed a 'Doctor Who' night to be broadcast and so I watched the programme. I was somewhat bemused and, I must admit, rather annoyed, as a not terribly good version of 'Muffin' was ushered from the Director General's office being told that he was no longer required as the BBC was making way for a new programme featuring a Time Travelling Doctor!

Muffin had been taken off air following Annette's death and consequently Muffin had not been seen on BBC television since early 1955. Dr Who started in 1960. I understood how frustrated and indignant Sally must have felt when Muffin was used inappropriately. Thankfully Sally did not feel that way when considering Muffin merchandising available exclusively to club members. To that end, I had suggested that we meet for lunch to discuss her thoughts and my ideas for the Club. Between our first and last meetings I had the pleasure to get to know Sally a little. Her wry sense of humour was coupled with a no-nonsense attitude to life and quite honestly, prior to our first meeting, I was extremely nervous about meeting her. That first meeting served to show me a little of Sally's attitude to life. She was a kind and generous person. This was apparent within moments of our meeting.

Sally and Derek arrived at our home in Essex. I suggested to my husband that we drive out to Epping, a town a few miles away where we had previously enjoyed a lovely meal in a local pub.

When offering the invitation to meet I had been quite restrained and had not asked if Muffin might be allowed to join us. I reminded myself that I was a mature (well, sort of) grown woman.

Upon arrival at the house, Sally came to the door and Derek went to the boot of their car, removed a bag and handed it to Sally. Once indoors with introductions completed, Sally opened the bag and carefully lifted the contents into view… Muffin!

Muffin was gently arranged on the floor next to Sally's chair and he received inspection from the family cat. The cat seemed faintly bemused at this creature that looked fairly familiar but smelt quite strange.

Sally lifted the control bar and, as if by magic, Muffin came to life. Sweep, the cat, eyed Muffin with suspicion but unlike the donkeys Sally describes later he stood his ground. After more judicious sniffing and standing at the same eye

Myself with Sally and Muffin.
Courtesy Ron Hasler

Sweep the cat meets Mufin the Mule.
Courtesy Ron Hasler

level Sweep the cat and Muffin the Mule touched noses. After this greeting Sweep stalked out of the room and Sally worked Muffin for me.

I was in heaven. How many of us actually get to meet our heroes especially in our living rooms! To ensure that I wasn't dreaming, Ron (my husband) took photographs of this momentous day.

Having had 'play-time' I was reminded that I had invited our guests to lunch and I drove to the pub.

On entering the restaurant section I was somewhat bemused to find Derek in deep conversation with a stranger. Sally grinned stating that we might have thought that we were the celebrities (having appeared on Collectors' Lot) but apparently not. After a short while Derek reappeared and explained that the person with whom he had been conversing 'was an old client'. Derek had been a Probation Officer and, unbeknown to me, some years earlier, he had been based at a courthouse just a short distance from our home.

Digressing slightly, I later discovered two other facts of interest – well, to me, at least. Firstly, I shared a birthday with Jan Bussell – 20th July (Ann Hogarth's was 19th July) and secondly, Ann Hogarth's maiden name was actually Jackson, as was mine.

(Again, I am indebted to Jane Phillips for her research and information relating to the Bussells.)

I understand that Ann had taken 'Hogarth' which was a family name because she was a direct descendant of William Hogarth, the 18th century artist.

It seems that I was fated to be drawn to Muffin!

Sally was extremely supportive of the Club and I believe (and hope) she enjoyed recalling and writing out her memories for the newsletter. I discovered that one of our members was a professional model-maker and that he had had prior contact with Sally. She knew that his work was well regarded and consequently we asked him for ideas to develop Muffin-related items. In the meantime another Club member created an embroidered logo designed for sweat-shirts and polo shirts, again for Club members only. Sally's copyright and control of these items was maintained by me. The shirts were popular and Sally took possession of tiny versions for Thomas her grandson.

Shortly after our first meeting in 1999 Sally became a grandmother for the second time as her daughter Lucinda gave birth to daughter Lizzie, a sister for Thomas.

Sometime later, I thought that I should arrange a day so that Club members could have the same opportunity and privilege of meeting Sally and Muffin. Sally, who was (in my view) extremely unassuming about her talent and her heritage, expressed concern that she and Derek would not fit in as they did not collect anything. After a moment's astonished pause I responded that I considered that they had the ultimate collection, after all she owned Muffin and the other puppets. And so, in the spring of 2001, Sally, Derek and puppets were guests of honour at our first Club Day.

Club members are a diverse group. Amongst the membership are members of various professions –law, teaching, nursing; we have artists, professional puppeteers, archivists, and others who work in antiques and for the BBC. The one thing we have in common is a passionate interest and childhood memories of Muffin the Mule and his friends.

My favourite photograph of that afternoon is one that I never dared to show Sally. I regret that decision because I think it gave a real sense of Sally – the joy and enthusiasm that she had for her life and her work. It shows Sally

Photos: Adrienne Hasler

manipulating Muffin wearing her jacket. Her buttons are wrongly done up and she is using facial expressions to act out the movements with Muffin.

Thankfully, the day was dry as my house bulged at the seams with enthusiastic Club members. We took to the garden for Sally's demonstrations with the puppets. Peter the Pup, Gracc the Giraffe, Mr Peregrine Esquire, Kirri the Kiwi, Sally the Seal, Hubert the Hippo and, of course, Muffin, took their turn to perform in the garden. As if this was not delight enough Sally had one more treat in store for us. She had brought a very special little film. It was called Muffin and the Reluctant Carrot. As has been mentioned already this was the first animated version of Muffin and made by her father.

Club members of all ages enjoy meeting Muffin the Mule 2001. Janet and Muffin - courtesy Wendy Garvey. Amelia & Muffin - courtesy Vicki Neville.

The film is in colour with the 'We Want Muffin' music played over the action. Spontaneous applause greeted its conclusion. We were indeed privileged as Sally told us that it had never been shown outside of the family before. Prior to our first Club Day Sally and I had discussed the idea of putting together her recollections and my photographs and to this end I'd asked Club members to bring items that I did not (yet) have, to be photographed. Sally now had a computer and we kept in touch by email and telephone. In May 2002 I was shocked to hear that Derek had been taken very ill. Thankfully, he made a good recovery and in August 2002 Ron and I planned another trip to Devon. I was looking forward to showing Sally a recent purchase I'd made. I had been contacted by someone who told me that he had a folder of paper/cardboard items which appeared to relate to Muffin the Mule. Would I be interested in buying them? Do I like to breathe? On inspection they appeared to be some of the artwork for The Reluctant Carrot and I bought them. I showed them to Sally and Derek. Sally identified the folder as her mother's creation, and the writing on the outside also. Sally and Derek werc as astonished as I. We believe that the young man who initially bought the folder saved all that is left of the artwork. He had been working in a little museum when the museum was offered various items by a 'house clearance' specialist. The museum curator selected a few items of educational value but didn't want Muffin and his carrot.

Thankfully, the young man thought Muffin was cute (no argument there!) and took the folder home. It languished in a drawer as he grew older, married and had a child. His house was getting cramped with his growing family and judicious clearing was being attempted hence the offer to sell the folder.

Muffin and his (Reluctant) Carrot are stored carefully. Sadly, Ann Hogarth's writing got a little smudged on being shown to Sally but is still legible. I feel sad that the rest of the artwork has gone, presumably for ever.

The artwork was Neville Main's and we can only think that, after his death, his house was cleared and his beautiful drawings and stories were overlooked and destroyed. Perhaps we are wrong and more is safely tucked away somewhere. I hope so.

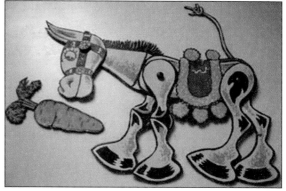

Muffin admiring his 'Reluctant carrot' - courtesy Adrienne Hasler. Puppet created by Neville Main for the colour animation by Jan Bussell 1956.

During that visit Sally told us that she hoped to have some wonderful news regarding Muffin. "Things were at an early stage and they had been there before" were her words but the message was clear. Someone was interested in developing Muffin for the 21st century. Sally was remaining calm and I did my best to emulate her but it did sound very exciting.

That visit to Sally was notable for another reason. Ron and I were staying in Sidmouth and had noticed that the Red Arrows were to give a flying display on the day of our visit. After a lovely day spent in the company of the McNally's we went into their garden and watched the display from the comfort of their lawn.

We expressed our appreciation to Sally and Derek for arranging this modest entertainment and they, with characteristic humility accepted our thanks! Unbeknown to us this would be our last opportunity to share such delights.

One piece of good news was that it was now (very) public knowledge that Muffin was to return to the BBC.

A company called Maverick Entertainment Group Ltd saw Muffin's potential in today's television and Sally negotiated a fixed term in which they could use Muffin's image in film and marketing. Sally would have input into how this was achieved and would be involved in the process at all levels.

When the news had broken in April 2003 the national press covered the story and, amazingly, the news was even featured in such diverse media as The Daily Times in Pakistan, The Education pages of the Guardian and a discussion on Radio 4!

Sally and I kept in email contact and I arranged another Club Day for September 2003. This date was later altered as Sally's family had arranged a surprise holiday. Mindful of the possible merchandising to come, Sally gave her permission for Paul Robbens (our model-maker club member), to create a multi-strung Muffin in the style of Neville Main's version. This puppet would only be available to members attending Club Day 2003. Sally would sign the control bars. Having seen the first puppet produced, I knew that Sally would be thrilled with Paul's work. He has the same high working standards and a passion for Muffin. A perfect combination.

As the new date, in October, got nearer I looked forward to seeing Sally and Derek again. A few days prior to the date I got a call from Sally to say that she was feeling wretched but still hoped to be with us.

Due to a disability, Derek cannot drive long distances and so we kept our fingers crossed that Sally would feel well enough to drive. On the eve of Club Day Derek rang to say that Sally could not attend. I was extremely concerned as I felt that Sally would have crawled to a prior commitment so as to not let people down.

It was too late to cancel and, as luck would have it, Jane Phillips was to attend her first Club Day. I tried unsuccessfully to contact her. Armed with my little 'Robbens' Muffin puppet I gathered up food and other necessary items plus my band of lovely volunteers and set forth to do my best to retrieve the event. I need not have worried. As I'd hoped, everyone was more concerned for Sally than themselves and Jane (and mini Muffin) proved to be a star.

Jane Phillips takes control of a 'mini' Muffin puppet. Jane was formerly apprenticed to Ann Hogarth and subsequently Lecturer on Puppet Theatre at Cardiff College of Music and Drama and Director of 'The Caricature Theatre Ltd' located in Cardiff, Wales. Photo: Adrienne Hasler.

Mini Mufin puppet: created by Paul Robbens (PRSFX) exclusively for members attending Club Day 2003. Photo: courtesy Adrienne Walsh.

I think (and hope) that Jane's success made it easier for Sally to accept that she had not let us down. Unfortunately, the same could not be said for the Post Office. Paul sent the puppet control bars to Sally in early January 2004 but she reported their non-arrival. Eventually, a second batch turned up in early February and Sally signed them. Sadly, mine is unsigned, I already had my puppet and didn't think to ask Sally to sign it. Sally's emails continued to be light-hearted and amusing. An abortive attempt at her section of the book provoked laughter not annoyance. She noted that she had been told that she had gall-stones and Derek had suggested they be utilized as Christmas presents! She greeted the news that our youngest son had moved to Plymouth by telling me that would give us even more reason to visit her! Frankly, I had refrained from frequent visits as I felt that I might very easily outstay my welcome! Visiting the McNally household was a great joy and great joys should be treated with respect.

At the end of March 2004, I received an email from Sally with the shocking news that she had cancer. In typically understated fashion she noted that 'We've been having a few dramas of our own down here'.

Our remaining correspondence was conducted by telephone, prior to and following her hospital stay. As a former nurse I was well aware that Sally's illness was terminal and she knew that this was so. I arranged a visit in early May. My daughter, Helen, and I would visit her brother in Plymouth and if Sally was up to receiving visitors we would visit her. On 13th May 2004, Helen and I visited Sally and Derek. Sally showed me the schedule provided by Maverick Entertainment for marketing and promotional events. It was quite clear that both Sally and I were well aware that it was unlikely that Sally would be able to attend any of these events but the conversation was positive and forward looking. Derek and Sally allowed Helen and me to watch the first portion of animation although they wickedly pointed out that they knew how it ended and we didn't! Sally described the 'treats' that the local hospice offered those in its care and her plans to explore other treatment opportunities.

Her daughter and son-in-law arrived with their two young children and it became apparent just how weak she was. Her granddaughter Lizzie asked Sally to work Muffin but Sally replied that she would have to wait until later. As Helen and I left, I knew that it was almost certainly the last time that I would see Sally. She died just seventeen days later, at home with Derek and William by her side.

In 2004, Sally had learnt that her son and daughter-in-law were to become parents for the first time. Sadly, she died before her new granddaughter, Georgina, arrived in December 2004.

In his UNIMA obituary Jan Bussell is quoted to have once said 'Our main wish is to evoke a joy in ourselves and in our public'. In my opinion, Sally's desire to find a way of bringing Muffin to a generation of 21st century children fulfils that wish. I'm sure that her parents would have been as proud of her as are her husband and children.

I feel honoured to have known Sally and regret that it was for so short a time. For those of you who never met Sally, I hope that the following newsletter contributions give you an insight into this generous lady.

Sally - courtesy Sue Johnson.

Far-Flung Muffins

One puppet I was responsible for finding was a Russian Muffin! Well, his photograph and location…

In his book, 'Through Wooden Eyes', Jan Bussell describes meeting the distinguished Russian puppeteer, Sergei Obratsov. Obratsov began his career as a puppeteer in 1931. Coincidentally, this was the same year that Jan Bussell was working as an apprentice in the London Marionette Theatre and he was part of the performance of "The Man with the Flower in his Mouth", one of the first puppet productions transmitted by John Logie Baird's experimental medium of television.

Obratsov was a legendary figure in the field of puppet theatre and it was on a visit to London that the Bussells eventually met him. Previous attempts at contact had failed, not surprising given that this was the era of the 'cold war'.

Obratsov's skill and the enjoyment derived by all parties is well documented in Jan Bussell's book, including the fact that he was extraordinarily impressed by Muffin (well, aren't we all!). This compliment resulted in a gift to Obratsov of an exact replica of Muffin for him to take back to his puppet museum in Moscow. The museum was to be gifted to the state after his death. And so Moscow Muffin sat quietly in his museum case until a day in February 2003 when I received an email from a young Russian woman who was to write a short biography article about Ann Hogarth!

The writer was intrigued when I sent her some photographs of a Muffin book in Russian. This was The Red Muffin Book (well, what else could it be!) and I gave her some information about Muffin on television. Her response was as follows: "On the whole the idea of a puppet 'starring' in a TV show (especially with all the production of different items organised around it) is not something really Soviet-like. But then if you think about it…We still have that one children show that survived from the Soviet times ("Good

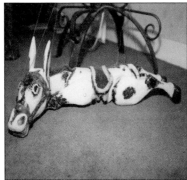

The Obratsov gift Muffin the Mule and his original self. Photograph from Russia, courtesy of Elina Absalyamova. Original Muffin, courtesy of Derek McNally.

Night, The little ones"). There's a human presenter and some puppets (a well-behaved hare, a naughty piglet, a clever dog, etc) - the tradition seems to go back to Muffin shows…."

After a few days of further correspondence, I asked the writer if she could locate the Obratsov Muffin. One phone call to the museum confirmed he was still in residence and looking particularly fine. So the next request was for a photograph.

This duly arrived and in June 2003 members of the Collectors' Club got their first view of Moscow Muffin. Actually, Sally had never seen him either. Rumour has it that there is another replica in New York, but thus far I have not located him.

More About Muffin

Between the years of 1946 to 1954, Muffin and his friends enchanted children by appearing on television on alternate Sunday evenings. As with many families in the 1950s, my parents were only persuaded to purchase a television to watch the coronation of Queen Elizabeth II.

Until the coronation in 1953 very few homes had a television. Indeed, some club members recall being invited to visit a neighbour's house with a television and the children sitting, as though in a theatre, in rows, to watch a very small screen. Sally recalled that after each broadcast a young neighbour would tell her mother what his friend Muffin had been doing the evening before...never appreciating how close he was to his hero!

Frustratingly for me, this meant that Muffin had been appearing on television for some six years before I ever got to see him!

The combined talents of Annette and Ann produced a brilliant and wonderful partnership. Of course the first impact of Annette Mills on Muffin's career had been selecting his name. As with many a star the choice of name can bring about great changes in circumstance, and so it was for Muffin...the mule formally known as 'the Kicking Mule' became Muffin the Mule and, unforeseen by the Bussells or by Annette Mills, his celebrity would continue into the 21st century.

Photograph taken from 'Illustrated' magazine, Nov 1947. A BBC Executive (and family) watching Muffin the Mule.

Alongside Muffin the Mule appeared his friends. These were Peregrine the Penguin, Louise the Lamb and Oswald the Ostrich with his constant companion, Willie the Worm. These puppets were created and made by Jan Bussell. Later came Katy the Kangaroo, Kerri the Kiwi and Zebbie the Zebra - made by Stanley Maile (originally puppet designer for the London Marionette Theatre and subsequently with the Hogarth Puppets) and created for the Hogarth Puppet Theatre tours of Australia, New Zealand and South Africa.

Other puppets, for example Peter the Pup and Monty the Monkey, were either taken from the current collection of Hogarth Puppets or were created for a particular story. I understand from Sally that Peter the Pup was made in the

Muffin and friends with Ann Hogarth and Jan Bussell.
Photograph courtesy of Derek McNally.

most unusual and dreadful conditions - the trenches of the First World War. Although Peter is very fragile, the Collectors' Club members have been privileged to meet him. Peter was initially made for Edith Lancaster (another leading puppeteer of the time) but she loaned him to Jan Bussell and eventually sold him to the Hogarth Puppet Theatre.

For more detail of this time, the reader would enjoy Jan Bussell's book 'Puppet's Progress'. Sally also recalled that some of her own toys were seconded to join Muffin in his considerable adventures.

Alongside the television programmes Jan Bussell produced Muffin films in order that the country's children would not have to forgo their fortnightly treat of seeing Muffin should he actually be touring!

Muffin and his fellow puppets travelled extensively filling halls, cinemas, theatres and tents throughout both the UK and abroad and The Hogarth Theatre caravan was a sight to gladden the hearts of children everywhere.

At the first Collectors' Club members meeting in 2001, Sally brought 'Muffin

Photograph taken from 'Puppets Progress'. Author, Jan Bussell 1956. Photo courtesy Corporation of London, London Metropolitan Archives.

and the Reluctant Carrot' to show members. This was a great privilege as the film had never been seen outside the family. At the time no one knew that, in just a few short years, a new generation of Muffin programmes would be created with animation. What we did know was that the film was charming. It was amusing and left viewers in no doubt that Muffin could have continued in this format if Jan Bussell had been encouraged to take his Mule in a new direction. Regrettably for 1950s and 1960s children this was not the time.

After retiring to Devon the Bussell's involvement with international puppet companies and their scholarship of puppetry meant that they continued to enjoy the company of puppeteers from many nations. They set up their exhibition of puppets and continued to give talks and demonstrations to fellow professionals.

After Jan Bussell's death in 1985, Ann Hogarth was frequently asked to 'work' Muffin for a variety of BBC occasions. Having moved nearer to Sally and Derek, some of these requests involved Sally supporting her mother and later you can read Sally's account of some of these times. Ann Hogarth died in 1993 and Muffin moved house - to live with Sally and Derek McNally in their lovely home in a village in Devon.

Muffin did not settle into retirement. Indeed, I suspect that he was busier than ever as Sally was asked to speak to groups from the Women's Institute, The British Legion and the various charities in which she was involved. Sally was also called upon to speak on a variety of radio programmes about Muffin, and appeared on Channel 4's Collectors' Lot programme in 1998. She also appeared in the Radio Times magazine with Muffin in August 2001. Muffin appeared on the front cover.

Muffin the Mule

Muffin the Mule Collectors' Club

I became involved with Muffin by accident. Until 1996 the only knowledge I had of Annette Mills was a very vague childhood memory of 'the lady with Muffin the Mule'. In fact, my recollection of Muffin was quite hazy too! A chance encounter was to change that.

New Year's Day 1996 saw my husband and me taking our usual walk to a local 'antiques fair'. By coincidence the fair (and I) live only a dozen miles from North Weald from which, as I later discovered, Annette Mills had the accident which would eventually lead to her partnership with Muffin the Mule. She was on her way home from entertaining the troops at North Weald Air Force base when her vehicle was in collision with another in the 'black out'.

Generally, I rarely find anything to bring home from antique fairs (often to the relief of my husband), but this time, not only did I bring home something special, but I opened a new dimension to my life too - I had found a shabby tail-less little 'Moko' Muffin the Mule puppet.

I bought the little puppet. This triggered the memory from my childhood and I resolved to find out more about Muffin and his companion. My research led me to discover some basic facts, and as my collection grew so did my quest for knowledge about Muffin. This eventually led to the creation of the Collectors' Club and my friendship with Sally McNally.

I had watched with considerable envy the interview with Sally (and Muffin) on her appearance on Collectors' Lot. Pressure from my husband to contact Collectors' Lot about my own collection led to my own appearance on the programme in 1999. Reaction to the programme from a puppet dealer was that he knew of several like-minded souls interested in Muffin, and why didn't I set up a club? Having taken the advice of the puppet dealer I decided to set up the club in the summer of 1999. Having done a little research I discovered that in the 1950s there was a Muffin Club which was promoted by the weekly publication, *TV Comic*. Further probing revealed that this club raised money to provide televisions and radios for children in long-term hospital care or suffering with chronic illness at home. I wrote to Channel 4 and asked if they might contact Sally on my behalf. They did so and one memorable afternoon the telephone rang. On answering it I was stunned to be greeted by Sally. Was Muffin listening in, I wondered?

I outlined my ideas for the club and was thrilled that Sally agreed not only to be Club President, but said that she would write about her recollections of life with the Hogarth Puppets. Her knowledge covered not only her participation as a young family member with recollections of the Hogarth Puppets on their travels but also her involvement in many of the television transmissions too. I later discovered that she had also developed some of the storylines to be found in the 'TV Comics' of the time.

Sally was extremely modest about her achievements and commitments and was somewhat surprised at the passion still generated by her little Mule.

Sally granted the necessary permission for the club to have its own (new) badge and I commissioned it. We used the image that was common to the notepaper used by Annette Mills, Ann Hogarth and by Sally herself.

Muffin is very much protected by copyright and I was extremely nervous being given responsibility for such a well-loved figure. So, taking a deep breath, I compiled the first newsletter - a very modest single A4 sheet in black and white and we were off!

In 1952 the original Muffin Club was set up by *TV Comic* (Beaverbrook Newspapers) and was promoted by a weekly page from Annette Mills. It included information on the activities of Muffin and his friends, his various travels to Australia and other far-flung countries and also about the club members themselves. Although I could not hope to emulate the success (or the professionalism) of the original club, I was keen to do my best. In setting up the Collectors' Club I decided that fundraising was one aspect that I could continue and so, with Sally's agreement and advice, I chose two charities to receive the net balance from our funds each year. These are 'Freshfields'- a charity, in Buxton Derbyshire, combining the care of re-homed donkeys with the care of children with special needs, and 'The Friends of Lyndon House' - a respite care home for children, situated in Solihull.

At the end of our first year (with thirty members) the club sent £120 to each of the charities.

In October 2000 I had a very special visit from Sally and Derek McNally. Sally and Derek were presumably used to the sight and sound of mature adults reverting to childhood when first introduced to Muffin, but they kindly hid any feelings of sympathy to my husband whilst I had a magical afternoon in the presence of a great equine actor. According to the recollections provided by Sally, I was in very good company. During Sir Winston Churchill's term as Prime Minister, Muffin and his human family were invited to entertain the Churchill family...here I was some forty plus years later, in a similarly privileged position. Minus a cigar, of course!

Sally generously gave of her support. She gave of her time (and Derek's) attending two Club Day events and, of course, Sally wrote for each newsletter. Since her death Derek has taken over this role.

When it came to our attention that one of the club members was a professional special effects and model maker, we were able to utilise his talents and provide club members with the opportunity to collect very special and exclusive Muffin the Mule models. In years to come I suspect these will be highly sought after. Indeed, I was amused (and somewhat bemused) to find a club member bidding on an auction website for an unauthorised copy of the club badge. The club website can be found at www.muffin-the-mule.com and we have 'links' with other sites relating to children's TV information and collecting. The website was set up and is maintained free of charge by my eldest son. The site offers some information about Muffin and the Club and is a contact point for people wishing to track down both memorabilia and knowledge. I have also been fortunate in receiving several lovely childhood recollections, for example, the tale of a pair of Muffin slippers! Unsurprisingly, the majority of members were children in the 1950s, but some are not. It is a testament to the remarkably enduring character that is Muffin that we have younger members and also requests to enrol children as members. As yet we do not have juvenile members, but given the likely response to the new Muffin programmes I suspect that a junior club will soon be under consideration.

I hoped that, as membership grew, we would be able to send an increased amount each year to both charities. Unfortunately, in common with many clubs, membership peaked and then stayed at a modest level of around fifty members - members for whose support I have been most grateful. Inevitably, I suppose, losing Sally was a blow to membership, but I hope that with fresh publicity about Muffin a few memories will be stirred and more members will join.

In spite of our grievous loss, by August 2004, our contributions had exceeded £2000 to each charity.

What next?

In 2003 Sally signed with Maverick Entertainment Group Ltd. Sally negotiated a fixed term in which they could use Muffin's image in film and marketing. With these rights they can use the characters in a variety of ways, for example, children's films, television programmes and related entertainment products and raise profit through the merchandising, licensing and video/DVD distribution. Such a contract gave Sally the opportunity to see Muffin developed for children of the 21st century and she was intimately involved in the selection of characters and early animation production.

At the time of writing, the new Muffin animation is still in production. A short promotional video has given me further indication of how the puppets have been recreated. I can report that the inherent characters remain intact, the animation is beautifully drawn and the humour is clearly demonstrated in the dialogue. The series looks set to allow yet another generation of fans to appreciate the magic of the little Mule. Sally and Derek's grandchildren will be able to enjoy watching Muffin's adventures whilst sitting with 'their' Muffin…a very special little wooden mule.

Courtesy of Maverick Entertainment Group Ltd
'Muffin the Mule' logo 2004

Collectors' Club Charities

Friends of Lyndon House

Lyndon House offers short stay care services for children with learning disabilities and a healthcare need.

The unit is part of the Solihull Healthcare NHS Trust but, as with many other units, relies on fundraising to help supplement the basic equipment required by such a unit.

The unit is staffed by specialist registered nurses and dedicated care staff, many of whom support the fund raising efforts of the Friends of Lyndon House. If you would like to have more information, please write to or email the club.

Play equipment bought with the assistance of Collectors' Club donations. Courtesy Emma Byrne

Freshfields is a unique charity in that it has a dual purpose.

It is responsible for rescuing donkeys from inappropriate situations and caring for them in a very special way.

The donkeys are cared for by children who, themselves, have special needs.

The children are allocated their own donkey for the duration of their stay and under supervision, the child cares for 'their' donkey.

Freshfields is the home of The Michael Elliott Trust (a registered charity) and is situated near Buxton in the heart of the beautiful Peak District National Park in Derbyshire.

Should you wish for further information about this charity, please visit their website at www.freshfields.org.uk

Muffin the Mule

Memories
as recalled by
the late
Sally McNally

Newsletter 1: October 1999

The origins of Muffin the Mule: by Sally McNally.

In 1933, my father, Jan Bussell, scribbled an outline of a mule on the back of an envelope and asked a Punch and Judy man to make him a kicking mule and a clown – presumably to be kicked. My parents had been touring with their show 'The Hogarth Puppets' for some time and had got very slick, consequently the running time was a little short and they needed a new act. This act with the clown being kicked and tossed around the stage proved very popular with the children but they actually found it rather boring to perform so it was eventually dropped. The clown was used in the puppet circus with various other acts but the mule was put on a shelf and forgotten.

In 1946, shortly after television had resumed transmission after the war, Annette Mills was appearing fairly regularly in Children's Hour. In those days there was only one TV channel and five o'clock on Sunday afternoon was set aside for the younger viewers, though in reality the whole family used to watch, TV being such a rarity then. Annette was performing her own songs at the grand piano and one day she observed to her producer that the top of her piano seemed such a waste of space surely something could be done with it to help bring her songs to life. My father was also a TV producer at that time and everyone knew of his passion for puppets so it was suggested that he should come up with an idea. Subsequently, Annette came to our house and met my mother, Ann Hogarth, and she asked if they could make puppets to illustrate her songs – no, they replied but perhaps you could write songs to illustrate our puppets! She agreed to this and my parents showed her all the puppets that they weren't actually using at that time. Annette immediately chose the mule and named him Muffin – then she also selected the clown and called him Crumpet. My mother sat down and wrote a twelve minute script, Annette wrote the songs, including, of course, the signature tune 'We want Muffin' and they made their first appearance together the following Sunday.

The programme was an instant success and Muffin started getting fan-mail immediately. Sadly though, Crumpet was not popular so he returned to the Hogarth Puppet Circus, re-assumed his name of Tickler and continued to amuse people with his antics in the 'live' theatre. Muffin went from strength to strength and various characters were soon added. My father made Peregrine the Penguin, Louise the Lamb and Oswald the Ostrich, who came with his best friend Willy the Worm, and so on. Peter the Pup and Sally the Seal came from stock and one or two such as Morris and Doris the Field-mice, from my toy cupboard!

They soon got into a routine, my mother would think of a plot and write the script, Annette would compose two or three new songs, each 'friend' also had his very own song, and then they would meet and rehearse. Later they abandoned the rehearsal time and did it all on the telephone. Every other Sunday morning they met at the TV studio for a camera rehearsal, a rather slow and laborious process, later there would be a straight run through of the whole programme before the live transmission at five o'clock.

Their success was amazing, and totally unexpected, no-one quite knew how to handle it. People appeared from everywhere asking for permissions to use Muffin and his friends in all manner of ways, from rides in the fair-ground to soaps, curtain materials, comics and so on. There were even questions asked about him in the House and the BBC claimed him as the first star to be made by TV. When you remember that until the Coronation in 1953 there were only about five hundred TV sets in the country, the impact that Muffin made is quite extraordinary.

Sadly, in 1954, after a short illness and at the early age of sixty, Annette died; they were at the height of their fame. However, Muffin continued, not so much on television but in the theatre with the Hogarth Puppets. They toured the world and Muffin continued to entrance audiences wherever he performed.

Ann Hogarth died in 1994 and Muffin now lives in Devon with me. Despite being in semi-retirement - after all he is sixty-six years old – he still makes personal appearances. It is surprising how often he pops up in various TV programmes, and there is even talk of a new series…….!

Newsletter 2: Feb 2000

Adventures with Muffin the Mule:

When Muffin the Mule first appeared on television all programmes were transmitted "live" from Alexandra Palace, and on high days and holidays I was allowed to accompany my mother to the studios. It was all rather exciting and glamorous in those days – it was such a new medium that everyone felt very involved and there was a tremendous atmosphere of camaraderie throughout. Children's Hour was shown at 5.00pm every Sunday and Muffin appeared fortnightly as part of that hour. The entire programme would be televised from one large studio, which was crammed full of paraphernalia, and at first glance seemed totally chaotic.

Cameras, lights and sound equipment, all with huge cables which trailed across the floor in coils, occupied the central area, all around the sides would be the scenery and sets needed for that programme's transmission. The Muffin set, which comprised of Annette's piano pushed against a mock wall, would occupy one space – next to them would be something quite different – possibly someone teaching the children how to make a kite, whilst across the other side of the studio there might be two or three sets where they would be busy rehearsing the "classic serial" (this always seemed to be about Nelson dying in Captain Hardy's arms). The very large and cumbersome cameras would be pushed back and forth across the central area as required. The producer sat in a glass-fronted box above the studio and conducted the proceedings talking through headphones to the studio manager, cameramen, lighting and sound engineers. You had to be very quiet once the programme began as the microphone could pick up any stray sounds, you also had to be careful how you moved in case you cast a shadow or, worst of all, walked in front of a camera when it was in operation. Once the programme had begun, the huge studio doors were closed and no-one was allowed in or out.

Occasionally Muffin appeared in Outside Broadcasts which were great fun. I remember one extraordinary day when the transmission was from the Children's Corner at London Zoo. Annette's piano had arrived safely, but nobody had thought about where my mother was going to stand to manipulate the puppets. Eventually a rather precarious platform was contrived using a plank between two elephant mounting blocks! This was quite successful but the mountain goats were a problem we found very hard to solve. Not only were they so agile that they leapt up everywhere and we were constantly falling over them but they also ate anything and everything, including the script and we needed to be very alert for the safety of Muffin and his friends!

Animals, though, are quite often a bit wary of the puppets, what are these strange things that move by themselves but have no smell? This happened in a comparatively recent programme at Pebble Mill when Muffin was asked to help promote Freshfields – a marvellous charity which combines giving handicapped children a unique holiday with rescuing donkeys – donkeys are such placid creatures that they make ideal friends for children with problems – placid that is until they met Muffin when they fled rapidly out of the studio with their handlers in hot pursuit!

This does not seem to be the case with birds, however. Whilst touring in New Zealand, we took Kirri the Kiwi with us and I have some nice film of him scratching around in the undergrowth with two real kiwis who take no notice of him whatsoever. Kirri's introduction to the programme did cause a few problems though. He was originally made at the request of the New Zealand government – we had returned from a tour of Australia with Katy the Kangaroo and they wanted to get in on the act – the problem was that the staple diet of a kiwi is worms. Some of you may remember that Oswald the Ostrich's best friend is Willie the Worm…….eventually, with the help of a special song from Annette, Kirri was taught to eat spaghetti instead and so all was well in Muffin's world.

Newsletter 3: June 2000

It is lovely to be making a third contribution to the Muffin the Mule Collectors' Club Newsletter and I understand that the membership is growing. My mother and Annette would have been so delighted.

This time I thought I would tell you about a more up-to-date happening. BBC TV outgrew their studios at Alexandra Palace fairly quickly and new premises were obtained at the old Rank Studios in Lime Grove, Shepherd's Bush.

Mrs Attlee (the then Prime Minister's wife), a Bishop (I can't remember which one, but there was a lot of correspondence as to whether he should wear his gaiters or not) and Muffin were asked to officiate at the opening ceremony. To be honest I don't recall the occasion at all, I must have been away at school - however, when some forty plus years later the BBC left the Lime Grove Studios the 'Late, Late Show' on BBC2 decided to do a documentary about it all. Muffin was the only 'opening' participant still around so it was decided that he should help with the closing programme.

My mother had by then turned eighty but she was very determined and we had a couple of fun days filming. The studios were literally being taken to pieces around us. The building was to be demolished and I suppose everything that could be salvaged was being taken. The producer's idea was for Muffin to "wander" through the empty studios, discovering props and bits of scenery which reminded him of various programmes and series that had been made there. For example Muffin found the caption used for the sports programme - then a clip of a footballer scoring a magnificent goal was shown - followed by a shot of Muffin leaping around in delight having kicked an equally fantastic goal! Mother worked the Mule beautifully, but when it came to scrambling around in a skip so that Muffin could be seen salvaging an old 'spinometer' (as used by Peter Snow for election forecasts) he was firmly handed over to me - she drew the line as well, at being shut in a cupboard so that Muffin could be 'discovered' browsing over old film.

She also objected to being involved with a shot of Muffin walking almost the length of the studio which meant that the puppeteer (now me) had to stand on the camera dolly with the cameraman filming between their legs - no, don't try to imagine it - it was not one of my better moments, but I managed.

As my father used to say, to my mother's intense annoyance "Oh, Muffin's an easy puppet to work. My wife does him"!

The closing of the Lime Grove studios was just one of a surprising number of appearances that Muffin has done in recent years. I am also asked, quite frequently, to allow him to appear in various TV programmes. He is often used to 'date' things, though the producers are sometimes way out in their time scale and he has been asked to indicate the nineteen fifties, sixties and even the seventies, and who am I to correct them? But the number of people who are quite adamant that they remember watching Muffin and who I feel sure would have been far too young is extraordinary. Only the other evening a gentleman was quite positive that I must have got the dates wrong as he said he distinctly recalled sitting with his children, watching the Muffin programmes, but according to me, Annette had died and the series ended before they were born. It is certainly true that Muffin found and has kept a place in many people's hearts.

Newsletter 4: Oct 2000

Sorting through some papers the other day I came across a few items I thought might be of interest to the Muffin Collectors' Club. Firstly a piece of Muffin wallpaper, then a few copies of the TV Times -featuring the mule, of course - then two splendid menus for the Cumberland Hotel.

Their Children's Christmas Gala Luncheon: 1951 and 1952. Children: 13/6d. Adults: 21/-.

The 1951 menu reads as follows:-

1952 Menu

Clear Soup Muffin
Creme Peregrine
Turkey Morris & Doris
Brussels Sprouts Peter
Potatoes Louise
Christmas Pudding Grace
Mince Pie Oswald
Yuletide Log Willie

MEMORIES - SALLY Mc NALLY

There is a nice cut out picture of Muffin on the front and inside he is seen on his piano top having a picnic with all his friends. The 1952 menu is much more elaborate. It is printed on stiffish cardboard, which is half folded. On one side is a TV set (a typical fifties one) with a picture of Muffin, by turning the card wheel at the side you can change the TV picture to show different 'friends' - there are six poses in all. I think Annette and my mother did about three of these Christmas Galas.

My father and I weren't overly delighted with these 'jauntings' as it meant that mother was out for the best part of Christmas Day, and since we were usually working in various different shows over that period our own celebrations took, of necessity, a back seat. I have copied a Christmas card that Stanley Maile (who made several of Muffin's friends) sent to us at that time!

But Christmas was always a busy period. Twice we were asked to Chequers, the Prime Minister's residence, to entertain the Churchill family. That was rather exciting for various reasons.

Firstly, we had to negotiate the sentries at the gate, which made us feel very important. Secondly, I was, at the age of thirteen and used to wartime austerity, very impressed with the lavishness of everything. Wall to wall carpets, central heating, and a Christmas tree in the dining hall which reached almost to the ceiling, past the galleried landing.

Care had to be taken to position the great man's chair so that he could get the best view, his cigars and brandy placed carefully on a convenient table was all of the utmost importance.

The family just gathered around him haphazardly for the show and we spent most of the time trying to hear the non-stop comments that he made. Later, when we were packing things away, water started to drip steadily through the ceiling.

'Oh those girls', sighed Mrs Churchill, 'They've left the bath water running again!'

Annette and Ann, together with Muffin, of course, were also asked to do Special Days at big stores, again usually around the Christmas period, and I have photos of them being literally mobbed by crowds of fans. These days it is difficult to appreciate the impact that Muffin had. I remember one big shop in London staged a moving tableau of Muffin and his friends as part of their Christmas decorations, this caused such large crowds that they obstructed the pavement and the police limited the times for the puppets to be animated.

When the original Muffin Club was formed one of its purposes was to raise money to buy TV sets for children in hospital. These are not needed these days, TV sets are everywhere. So, when Adrienne asked me if I could suggest a worthy cause to receive a donation, I thought of the Donkey Sanctuary in Derbyshire.

Muffin is already a patron and the basic idea of bringing donkeys and children together so that they both benefit seems a brilliant concept - and it works - they seem to develop a 'special' relationship. John Stirling, who runs the sanctuary with his wife Annie, was delighted with the cheque. Their website, if you want to know more about them is:
HTTP://WWW.FRESHFIELDS.ORG.UK/

So, from Muffin and me, the Donkeys and the children who go to visit them a very big "Thank you".

Newsletter 5: Feb 2001

The first time Muffin, together with the Hogarth Puppets, my parents and I went to Australia was in 1951. I was fifteen and it was very exciting. In those days Australia was much more inaccessible then than it is now - for a start no one would even consider flying, you travelled by sea and the journey took five weeks! My parents were rather reluctant at first, largely as it would mean being away for such a long time but the Rayner Sisters who ran the Australian Children's Theatre were very persuasive and when the British Council offered a grant for our fares and expenses they felt they had to agree. Also they had already made some films to be shown on TV so these could be used in our absence.

There was not a very large gathering to see us off on the bitterly cold morning that we sailed. A few pressmen, one or two friends and, of course, Annette Mills, who was staying behind to continue with her Prudence Kitten series, both she and my mother were given large and beautiful bouquets - and I was rather cross at being asked to carry a special one for Muffin - made out of carrots - well, these things matter when you're a teenager!

We sailed on the P. & O. ship S.S. Strathnaver, in great luxury, travelling first class. I had a wonderful time, a glorious holiday. Once we had sailed into the warmer waters we played deck games all day and danced all night, and we called in at fascinating ports as an added bonus. Naples, Algiers and then through the Suez Canal and onto India. It was on this next part of the voyage, about eight days without seeing land, that we were asked to do a show on board. We had already done some filming, taking pictures of Muffin looking at the sea, watching flying fish, enjoying the 'crossing of the line' ceremony as we crossed the Equator and also arriving at various places where he could be seen riding elephants (with a very worried looking Ann Hogarth) or watching the 'gelli gelli' man who came on board as we went through the Suez and did amazing conjuring with tiny chicks.

These films, rather like travelogues, were sent with accompanying letters for Annette to read on TV. However a show was a more serious undertaking, the boxes full of puppets had to be got out of the hold, plus their props and scenery. We were all thrilled and touched to find a note had been pushed in to one of our big boxes, it said - 'Muffin and Co. - Good Luck - from the Tilbury Dockers'.

The show on board was a great success, however we made an interesting discovery, marionettes [string puppets] tend to take on a life of their own when operated on a ship at sea, no matter how slight the swell.

We landed at Melbourne and were given a marvellous welcome. A group of Australian puppeteers had come to meet us and the Rayner sisters, Joan and Betty, came with our living accommodation for the next nine months, a caravan which we towed with a three-ton truck that had our name emblazoned on the side. Our great Australian adventure had begun!

We were breaking entirely new ground; television hadn't arrived in Australia so hardly anyone knew of Muffin and his friends. Nonetheless, they took him, along with all our other puppets, straight to their hearts and we had a brilliant reception. After an exciting week or so in Melbourne we set out on tour and soon got into a routine. We would drive to the venue, which could be as far as 100 miles, find the school or hall were we were performing, meet the local organiser and set up our puppet theatre. This took about two hours, unless it was a difficult 'get in', for example, we sometimes had to carry all the gear up two flights of stairs, when it obviously took much longer. The following day we would do the shows, usually three, the first one was sometimes at 9.30 am, which is a ghastly time to put on a performance. There was nearly always a reception laid on after the last performance when we would be introduced to the local dignitaries and Mother and I quickly discovered, rather to our dismay, that we were expected to wear gloves and hats for this. Then we had to dismantle our theatre and pack it all up ready for the next day.

Back home in England the films - which have now been made into the videos - were being enjoyed by Muffin's fans, as were the travelogues. I have several strange bits of the travelogue film. I am not sure if it was ever shown except as a series of clips, one day I must dig it out and have a look.

I look forward to introducing you to Muffin in the flesh - or should I say wood.

Newsletter 6: June 2001

It was lovely to meet so many of you earlier this year. Adrienne certainly organised a splendid day - even the weather co-operated (at least it wasn't raining) and everyone seemed to get on together extremely well, though Derek and I felt that we stuck out a bit as neither of us collect anything! But it really was a very successful 'do' and I hope that the Hasler family weren't too exhausted afterwards. Many, many thanks to Adrienne, and also to Ron, for putting up with us all.

As the Muffin programmes were all transmitted 'live' I am frequently asked by enthusiastic reporters what went wrong during the programmes - what funny incidents happened as a result etc. Well, of course, things did go wrong, strings broke or got tangled - I remember one occasion when Annette had an extremely bad coughing fit which left her almost speechless for what seemed a lifetime, though in reality it probably lasted only seconds - but my parents, who were very professional, thought of these various mishaps as disasters rather than funny episodes. So, here is the story that I usually fall back on - which I think is rather charming and doesn't involve any mishaps, but my apologies if you have heard me relate it before.

In the early fifties it was usual for a family to all get together on Sunday afternoons, to have tea and watch television. In those days the "blanket TV ' we have now didn't exist, the hours of transmission were very limited, evening viewing began at about 7pm, lasting until about 10.30pm and for children there was one hour on Sunday afternoons from 5-6pm. The telly in the corner was treated with great respect, not many people had one and as a consequence it became a bit of a status symbol; so when it went wrong it was rather a disaster. On this particular occasion not only did the set go wrong but it also started to emit an unpleasant smell! The TV man was sent for and the whole family gathered to watch him repair their magic box. When the back of the set had been carefully removed the cause of the problem was immediately obvious as, to their horror, surprise and embarrassment, out fell mouldy bread, bits of salad and rotten carrots. On investigation it was discovered that the young daughter of the house was so concerned that Muffin always missed his tea on Sundays that she had been 'feeding' him through the small slits at the back! I think it shows what a very considerate child she must have been.

I had quite a 'find' last week. Shortly after Christmas I was interviewed on our local Radio Station, Radio Devon. During the broadcast I mentioned that my mother had lent some puppets to someone on what she liked to refer to as permanent loan, whatever that may mean. Anyhow I said that I had lost track of where a lot of these puppets had ended up. At the beginning of May I had a phone call to say that the Devon Resource Centre had got some, would I like to go and see them and check out what they had. Well, of course, I was more than pleased to do this and Derek and I duly set out for Torquay. The puppets they had were all from the Hogarth Puppet Theatre show, being looked after splendidly, and used to send out to schools (on request) to back up various subjects that were currently being taught. They were rather mystified though by a rather strange animal - mule shaped, its body made out of a National Dried Milk tin - that was hanging in the corner. I was able to tell them that it was Muffin in his suit of armour, dating from a programme called 'St Muffin and the Dragon', I am not sure, but I imagine Peregrine must have been the Dragon!

Have a lovely summer everyone, good wishes to all, from Muffin and me.

Newsletter 7: October 2001

It was lovely to meet up with some Club members at the Collectors' Fair in Reading, I think the organisers were pleased with the day and hope to make it an annual event. Adrienne seemed happy also - she got a few new members - but I expect she will tell you all about it elsewhere in this Newsletter.

I enjoyed it as it took away the nasty taste left in my mouth by the Channel 4 programme '100 Best Kids Shows' which went out on August Bank Holiday. Despite talking to me on the 'phone, they still managed to get their facts wrong, and fancy managing to devote quite a lot of footage to Muffin without mentioning Ann Hogarth once. Apart from anything else it is so rude.

Anyhow I have been promised a guarantee that if the programme is shown again, they will re-edit....... I wonder. I wish that researchers would do their work properly! However, Jack Tempest's article in the September issue of 'Collectables' cheered me up no end.

The summer seems to have flown by this year - probably just me getting older, but soon November will be here, traditionally the season of fog. Which reminds me of a strange occasion, way back in the days when we used to have those frightful 'pea-soupers'. For you young-things who don't remember them, they really did blanket out everything. Not only was it almost impossible to see, I once walked into a lamp-post, but it muffled sound as well and the overall effect was totally disorientating. We were on our way to the TV studios at Lime Grove, (at this point I must remind you that these programmes were going out live and punctuality was important) and although we had left very early to allow for travel problems, the journey was taking hours.

At one point we had joined a queue of cars following a bus. The conductor was walking in front with a flare to lead the way but we rather lost confidence in him when we discovered that he was taking us the wrong way round a roundabout! Eventually our driver just pulled up at a Tube station, this it seemed would be a more sensible way of travelling. Only one problem, there we were with all the puppets, props and general paraphernalia, and as we were expecting to travel by car they had been 'gathered' rather than packed. My mother had written a script all about bathing for this programme which, needless to say, called for some unusual items - the bath robes, flannels, towels etc., all in assorted sizes to fit everyone from worms to ostriches were fairly easily fitted into cardboard boxes, as were the ice bags (to put on your head whilst relaxing in a bath) but how do you pack a Turkish bath for a penguin, or even an ordinary bath for a mule come to that, especially one equipped with a geyser?

Pauline Jackson, Mother's super stage manager, was with us, and the three of us got some very odd looks as we struggled up and down stairs, along those seemingly endless underground passages and even managed to change trains with our strange assortment of luggage, before finally arriving, exhausted, at the studios. I don't remember there being time for a rehearsal, but mother had briefed us both fairly thoroughly on the train!

There is only one other similar occurrence that I can remember which happened when Muffin was appearing in the West End at the Cambridge Theatre. Muffin's Christmas Party was a revue for children and we gave morning and afternoon matinees. Again there was a thick 'pea-souper' fog and when the curtain went up there was less than half of the company present. The audience was also very small but we pressed on and by the time the curtain came down, some two hours later we had a full complement, both back stage and front of house.

By the way, have you ever tried to buy a pair of tennis shoes for an ostrich? I was a little surprised one day to hear Mother telling the shop assistant. 'Oswald has a short foot, but needs a rather wide fitting!'

Newsletter 8: February 2002

At this time of year it seems to me that winter is interminable! I long to sail off to some sunnier climes and think longingly of our tours to Australia, South Africa and New Zealand. I am sure I only remember the good parts, but overall it is the laughter that stands out most.

On that first tour we did to Australia in 1952 we really were trail blazers, living in a caravan which we towed behind our three-ton truck when travelling to the more remote places.

We quickly discovered the joys of the 'chip-eater' a sort of primitive geyser in which the water for a shower was heated by a handful of wood chips. We also got used to the lavish use of starch by the local laundries, our underwear took on a whole new lease of life! In those days very few of the roads outside the towns were bituminized and our itinerary would indicate what sort of road surface we would be encountering. We found there were several grades, ranging from 'Good Gravel' through 'corrugations' to, worst of all, 'corrugated Gravel with Pot holes', when we would be bouncing along, clinging to the sides of the van and at times literally hitting our heads on the ceiling. Our complaints to a garage attendant about this was met with a laconic 'Aye, you would', not very encouraging.

Sometimes, too, there was no sun, just interminable rain which not infrequently resulted in the gravel roads being washed away and a river to be negotiated. Then I would don my 'wellies' and walk in front - my parents said that if I totally disappeared they would have to find another route.

During these tours my father would take film of Muffin having various adventures, for example meeting a koala bear or a wombat. Occasionally my mother would rebel at his requests and I had to stand in as her substitute. One of these 'stand ins' was in Kalgoorlie in Western Australia, where we were given a tremendous welcome by the gold miners.

They had made a special miners hat for Muffin, complete with a lamp that switched on and off (and holes for his ears) and wished to make a special presentation. There is a tradition there that it would be bad luck if a woman went down a mine shaft but after quite a bit of discussion it was felt that descending the first 100 feet in the open-sided lift (that was the point when my mother opted out) would be all right. I duly donned a hard hat and with several large, burly miners was whisked away toward the bowels of the earth, where we paused for a while, then a swift jerk and we started our journey upwards again. When we arrived at the surface out we trooped, acting nonchalantly for the benefit of the camera, where these charming men knelt down and gently placed the beautifully made miners helmet on Muffin's head and presented him with a small bunch of carrots. They then took me off to teach me how to play 'two up', but that's another story!

Another occasion when I had to act as a stand-in was actually during the voyage to Australia.

One of the ports of call was Colombo, in Sri Lanka, or Ceylon, as it was called then. Somehow, I have no idea how, my father had managed to arrange for Muffin to have a ride on an elephant.

My mother was not amused, and even less so when it was discovered that in honour of the occasion, the elephant had been well bathed and was consequently dripping wet.

There was no question of a saddle or a houdah needless to say, but the keeper, mahout is far too grand a word, had thoughtfully provided sheets of newspaper for her to sit on, whilst astride the elephant. It goes without saying that I was the one who spent the rest of the day with newsprint indelibly etched on my dress and legs.

Nowadays, when travellers whizz around the world by air, they certainly miss out on local colour.

Newsletter 9: June 2002

As a result of Muffin appearing on the front cover of the Radio Times last year I was invited to a splendid party! Apparently every year a party is given solely for people who have been photographed for this. Since it was Muffin who had been in front of the camera and not me I did feel a bit of a fraud, but I tucked him under one arm and, I have to say, rather hesitantly at first (no extra tickets were allowed for husbands) I set off to the BAFTA centre in Piccadilly.

At first I thought I must have been mad to accept and lurked in a corner clutching my glass of wine rather nervously. But then I spotted the young man who is involved with the Wombles behaving in an equally 'I wish I wasn't here' manner and we teamed up. We decided that it was unlikely, certainly as far as I was concerned at any rate, that we would be invited to such a select gathering again so we would stare at the celebrities and make the most of the occasion. So I unpacked Muffin and we began to mingle. Muffin, as I should have known, proved a brilliant icebreaker. Alan Titchmarsh was very effusive, as was

Jonathan Ross, looking very resplendent in a suit of blue sequins. Ian Hislop was there, Gary Rhodes couldn't make it and I didn't quite have the courage to go and interrupt David Attenborough and Barry Norman in their very absorbing conversation! The climax of the party was Jonathan Ross presenting everyone with a framed picture of 'their' cover. Then out into the rain clutching a huge bag containing the picture and wielding aloft an equally huge emerald green umbrella, both stamped with Radio Times in large letters. A lovely evening to remember.

Last month I was surprised to get a phone call from Radio Newcastle. Would I do a phone interview about Muffin? A few years ago I had done one for the same chap, John Harle, for his afternoon show. Now he has a late night show and wanted to talk again. I was quite happy, though I did check to see what time he would phone me. Fortunately I was on fairly early in the show, at about 10.40pm. It is very odd to sit comfortably in your own sitting room, gossiping happily on the phone to a perfect stranger and wondering if anyone in the Newcastle area is either listening or the slightest bit interested! But I was able to plug the new DVD in case there was a listener.

Which brings me to the next bit… The DVD and/or video. Eight episodes being shown.

The packaging is splendid – with little gold Muffins on the actual disc. But I expect by the time you get this you will all have your own! Tomorrow, the 27th is D-Day, or rather M-Day*. (*See update) Our local bookshop is putting on a display, Muffin is in the window and whilst I was down there this morning quite a few people were stopping and pressing their noses against the window –'look there's Muffin the Mule', 'no, it can't be,' ' yes it is' sort of squeaks were floating around.

There are quite a few photographs, a mug, some wallpaper and various other items which I have dug out for the display. He is surrounded with Union Jacks as well, after all it is the Golden Jubilee and Prince Charles used to be a regular viewer.

 I feel I must add a line here to say how super the Newsletter is looking now. Adrienne is doing a marvellous job of editing (a blushing editor here) and I find the articles really interesting to read. Even if a bit disconcerting at times when I read about myself in another life!

Newsletter 10: October 2002

Firstly, a big 'Thank you' to Adrienne for giving us such a good day out on September 21st. Derek and I enjoyed ourselves very much, it is a real treat for me to get the puppets out and to see them being appreciated. I do hope the various photos all came out successfully! I also hope that Adrienne and Ron weren't too exhausted; they both worked so hard looking after us all.

It was in September 1996 (last century!) that the Royal Mail brought out its commemorative stamps for Children's TV. These, of course, included Muffin on the 20p stamp. It had all started about eighteen months previously when I got a phone call from Newell and Sorrell – who were going to do the design and art work – telling me that this was a possibility, wanting to know if I would have any objections and finally, swearing me to secrecy, saying that if the information was leaked in any way all negotiations would be stopped. It seemed rather similar to being awarded an MBE!

A few months later I was contacted again and asked for photos and once more sworn to silence. (I did tell Derek, otherwise I think I would have burst.) Then out of the blue a friend rang me to say that he had heard that Muffin was going to be on a stamp – was it true – I sort of waffled on the other end of the phone, but a few days later I got the confirmation, signed the agreement and it was all public knowledge. That was in January 1996. Newell & Sorrell sent me a copy of 'TV Years' an education pack on TV which took the form of a magazine and used the stamps as a starting point. This was sent out free to 23,000 primary school teachers in the UK. There were also Presentation Packs, Collectors Packs, Collectors Books, First Day Covers, Postcard Enlargement of Stamps, etc.etc. All quite exciting.

Then came the down side – Hansard reported a question asked in the House 'Lord Kennet asked what royalties will be paid and to whom for the use of Muffin the Mule on postage stamps.' The answer from Lord Fraser of Carmyllie was 'None'! Various other Labour MPs wanted to know why Muffin the Mule and other puppets were being commemorated and not William Morris, whose centenary it was - 'Stamps are about design and William Morris was one of the great designers..........It's like having Mickey Mouse instead of Winston Churchill' said Tony Benn. Other comments came 'I think it is astonishing these pygmies can elevate the contribution that Muffin the Mule has made to our history over that of an inspired visionary like William Morris'. 'This is unbelievable. It is just a completely ridiculous set of priorities and a huge insult', were a few of the comments. I must confess that although it seems a bit disloyal, in some ways I have to agree with them, but it is good that occasionally the powers that be can let down their hair!

Anyhow the BBC thoroughly enjoyed themselves – with double page spreads in the Radio Times and various other articles. I also had some fun as a result, being asked to do various radio interviews and also to take Muffin to Birmingham to appear in a special edition of 'Telly Addicts' with Noel Edmonds…. but that's another story.

My only regret is that neither Annette, nor my mother, were alive at the time – they would have been absolutely delighted, very proud and also highly amused by the whole episode. And just think what a brilliant programme mother would have written as a result.

Newsletter 11: February 2003

The Queen of Hearts

I do hope you all had a good Christmas and festive season. We enjoyed ours but it went on slightly longer than we had intended as our grandson had chickenpox for one of his presents – the first spot arrived on Christmas Day – and then, of course, his younger sister managed to catch it and the quarantine period seemed to drag on. Rather as the pantomime season does after the holidays are over.

Muffin did one 'proper' pantomime and a couple of Christmas shows. The 'panto' was 'The Queen of Hearts' at the Wimbledon Theatre. Two big names headed the bill apart from Muffin, both slightly past their sell-by date. Evelyn Laye, who must have been well into her fifties, but was a magnificent principle boy, with stunning legs, and Bobby Howes, a comedian, light entertainer, dancer etc. playing the Dame. I don't know if it was his first effort as Dame, but I believe it was his last! Enough said.

Also, we had Jimmy Wheeler, a brilliant comedian, taking the part of the King.

Quite how Muffin got involved in the story I'm not really sure, but the Tarts ended up on the top of Annette's piano doing a splendid dance with him – three of the six were operated by me – to one of Annette's songs. 'Now here you see the magic tarts I stole them from the Knave of Hearts' and so on.

Poor Annette used to suffer from migraine quite badly and I suppose the stress of an opening night would have helped to bring on one of these horrible headaches, anyhow she was unable to make the First Night…

So my unhappy father was roped in as her understudy. He could only play one tune on the piano, so he and my mother rewrote the words to fit that. For the other songs the band played – although he had a lot of acting experience he'd never worked with a band before, finally to add to his discomfort, the only costume they could find to fit him was doublet and hose!

However, Annette duly recovered and Pa was able to return to normal. The show was quite a success and when the pantomime season ended, it was decided to stay there a bit longer doing the whole performance.

The programme began with The Hogarth Puppet Orchestra followed by another item from their repertoire, then an interval, after which Annette appeared with Muffin and his friends, another brief interval before the final act, The Hogarth Puppet Circus. I think this was when Muffin first started appearing in the circus in the role of clown.

For example, there was a very good act in which two dogs, Blackberry and Snowdrop, tossed a ball to each other, catching it on their noses. Immediately afterwards, on came Muffin, wearing a ruff and with a large balloon balanced on his head. He pranced proudly across the stage but when he reached the opposite wing the balloon burst, Muffin collapsed and a very miserable mule made a dejected exit.

On another occasion he entered after the Strong Man, Mr Henry Mobbs, had just finished struggling with his exceptionally heavy weights and to Mobbs's fury, helpfully carried them off the stage as if they weighed nothing. But, for his final entrance, in the circus, Muffin, wearing plumes, as befitted a star, followed on after Flash the Cowboy Rider on his horse Sparkle, and would take his bow to well deserved applause!

Shortly after this Muffin, Annette and the Hogarth Puppets were booked to star in variety at the Chelsea Palace, topping the bill with Alfred Marks, a very interesting experience…more next time.

Happy Easter everyone, don't eat too many chocolate muffins, now there's an advertising opportunity!

Newsletter 12: June 2003

'My name is Apricot' said the voice on the phone. 'Really', I replied, well, what else was there to say? 'Yes' she said firmly 'and I have been asked to invite you to bring Muffin the Mule to appear on the Channel 4 programme, Collectors'

MEMORIES - SALLY Mc NALLY

Lot'. That explained everything. Where else but in the entertainment industry would you find a girl called Apricot. I'm afraid I never had the courage to ask if it was her given name.

And so it was that a couple of months later Derek and I found ourselves staying in a very nice hotel in Evesham.

This happened a few years ago before, I think, Adrienne's memorable appearance. Though, of course, it was as a result of the Collectors' Lot programmes that the Muffin Collectors' Club was born.

Sue Cook was still presenting the series and this particular programme was being filmed in a house overlooking the River Avon. It was a specially designed house, circa 1920, with everything, including furniture to match, I recall rather strange turrets with circular staircases and rooms in unexpected places all done to maximise the spectacular views.

As we arrived a most attractive young girl came to greet us, 'I'm Apricot' she said, and turned to Derek 'I am looking after you today'. Derek went off very happily with his plum!

Sue Cook was absolutely charming and made everything very easy. I was very impressed with her efficiency. Muffin, on the other hand, was not on his best behaviour. Firstly, he broke a string, not a totally unusual happening but fiddly to mend especially under those circumstances. Next, the props man complained that 'your mule has a bad case of dandruff!' Well, if you must give him black velvet to dance on……. but I made a mental note to give him a fresh coat of paint before his next appearance.

There was another interesting occurrence. The cameraman had apparently learnt his trade on the Muffin programmes, way back in the early days, I think at Alexandra Palace. This was to be his final shoot before retirement, the wheel had come full circle, Muffin being his first and last shoot for television.

This was not the first time that Sue had met Muffin. My mother and I had travelled up to Birmingham for the lunch time programme that used to be presented daily from Pebble Mill. This was mainly to help promote the Freshfields Donkey Sanctuary which was just opening. On that occasion, Oswald the Ostrich and Peregrine the Penguin were allowed to accompany us.

Mother did a very good interview, she was a natural for these things and could talk fluently and amusingly with no apparent effort, I lurked in the background manipulating Oswald and Peregrine as required. All went well, until the three donkeys arrived. Sue Cook gave them a good build up and they were brought into the studio, Muffin bounced forward to greet them and the donkeys, well, all I can say is that they were horrified. They retreated very rapidly with their handlers in hot pursuit and their interview had to be conducted outside!

Freshfields Donkey Sanctuary is one of the two charities that Adrienne supports with donations from the Muffin Collectors' Club. Apart from the obvious connection, there is another rather obscure one. John Stirling, who runs the Sanctuary, is the son of a great friend of my parents. My father travelled around France in the twenties with an English speaking company, mostly performing 'The Importance of Being Ernest' which was run by John's grandfather!

There is a very nice photo of him aged about 4yrs sitting on Annette's piano talking to Muffin. Perhaps that's when his empathy with donkeys began. Have a lovely summer, everyone.

I look forward to seeing you at the next Muffin Day.

Newsletter 13: October 2003

Greetings from Sally

I was so very disappointed not to be at the 'Muffin Day' this year. Derek, I - and Muffin, of course - had all been looking forward to meeting up with old friends and making new ones. However, I gather from Adrienne, you all had a good time and that my great friend Jane Phillips saved the day, talking about her time with my parents and how she used to help with the Muffin shows.

Did she tell you about her puppet theatre, Caricature Theatre Company, which she ran for many years, writing, designing, making and producing much of the material herself.

Although based mostly in Wales she also toured extensively in the rest of the UK and Europe and had a lot of very well deserved success.

I gather that the 'day' was rounded off with Jane showing part of a video (that she had brought to give to me) about the life of Lotte Reiniger. Following on from that I thought this 'memory' of Lotte that I wrote a while ago might be of interest.

Lotte Reiniger made the most beautiful stop-action films with small, delicate, cut-out figures. I first met her when my father came home with this huge, and rather noisy Austrian lady. She looked down at me over her spectacles 'Ach', she said, 'So this is Sally' and I seemed to disappear into her voluminous bosom.

I sort of squeaked my reply through the mixed odours of garlic and large warm woman for one of her many necklaces had entwined itself painfully in my hair.

My mother plied her with coffee. 'Have you got the designs with you', my father asked. 'Ja', Lotte nodded, 'Zey are here'. She tapped her head 'zey are ver good'. She beamed happily at us all.

'There is the Card you asked for'. Pa indicated the sheets of Card on the table. 'Is there anything else you need?' Lotte rummaged in her capacious bag and eventually produced several pairs of scissors of varying sizes. 'I have everyzing – now, go all of you, I must vork'.

Later, I was sent to call her for lunch. I knocked timidly on the door 'Come' called Lotte, 'Come and look'. I opened the door and with a certain amount of difficulty I waded through the small mountain of paper clippings that surrounded her. 'See what old Lotte has done'. On the table were the most beautiful and imaginatively cut figures and scenery for Oscar Wilde's story of The Happy Prince. There was the square with statue rising high in the middle, a small group of awe-struck children gazed up at it, a young girl in a ball dress, every flounce clearly shown, tossed her head, her soldier lover sighed in the background, the match girl shivered in the snow and the poet starved in his garret, whilst over them all flew the swallow bringing the wealth from the Prince's statue. 'You see', said Lotte, 'I said zey were good'. And she was right.

* * * * * * * * * * *

The last few months have been quite exciting and interesting ones for me and my mule. The new series being produced by Maverick Entertainment is actually going ahead with the first films being aired in 2005. I have read some scripts and seen some of the cartoon sketches. Richard Ollive is doing the animation, I'm sure you must have seen his Tetley Tea Bag ads and the Ribena ones. Jimmy Hibbert is writing the scripts, he does quite a few children's programmes and wrote the award winning Christmas Special for Bob the Builder. So, Maverick has got a good team together. There are, of course, some more Muffin-type spin offs – Muffin has been asked to be Patron for a RSPCA project in Devon. This means that he and I will be going along to various venues during half-term week, helping to raise funds. The week culminates in a Halloween Festival, which includes a 'Jelly Welly' Race!

Greetings to all,

Sally

Newsletter 14: February 2004

I hope it is not too late to wish everyone a Happy New Year. The daffodils, crocus and snowdrops are out and one feels spring is just around the corner. We only had a suggestion of snow down here in Devon but I gather that it has pretty well cleared from the rest of the country now. We just have rain, rain and yet more rain and I understand that our reservoirs are all full again after the long hot summer of 2003. So, to cheer myself up here is a story about summer in 1950, when my parents were asked to do shows on Woolacombe Beach for the month of July.

Muffin was fairly well known at this time, though not in Devon as TV was only just arriving there, and also very few people had TV sets at that time.

My parents had a caravan theatre which they used to perform 'open air' shows in the London Parks and various other venues. It could be converted from a bedroom to a theatre very quickly and we had some lovely working holidays with it. One side was opened to reveal a puppet stage and the roof of the caravan lifted up to allow both space and light. There were special places to put the gramophone (no CD's, tapes etc. then), loudspeakers and mikes.

The idea was that they would sleep in the caravan at night and do shows on the beach during the day for the month of July.

A nice holiday, my father said, but I don't think my mother agreed!

My uncle collected me from school and he and I drove down to join them a few days after the season had begun. Lawrence and I each had rather small tents to sleep in – though I fared slightly better as apart from there being rather less of me than of him I didn't have to share mine with a huge sheep dog. Our camp site in the dunes also included Jack Whitehead, our stage manager, and his wife and two sons.

There was also a Belgian couple, photographers, who took photos on the beach and then developed them in the gents in the evening! (No digital or Polaroid cameras then.)

My parents gave three shows a day – three circuses. Hoopo's Circus, Tickler's Circus, and, of course, Muffin's Circus. Hoopo and Tickler were both clowns, but Tickler is of interest as he was the clown made at the same time as Muffin and whom Annette had christened 'Crumpet'. Following his rejection by the TV audience he returned to the Hogarth Puppets and re-invented himself!

He now appeared in five different guises – a stilt walker, sword swallower, juggler, as himself, dancing, and as a grand finale he split in two from head to toe and thus frightened a dissecting skeleton. Muffin also excelled himself, lifting weights with his tail amongst various other activities, though the highlight of 'his' show was his dance with a balloon which, needless to say, bursts – his dejection when this happened was marvellous to see – it occasioned one of the best laughs of the show except from the very young who sometimes came round with their balloons to cheer Muffin up.

I had a marvellous summer, helping with the shows and selling Muffin balloons and brooches afterwards. I also managed to learn how to surf, no, not the clever stuff just belly boarding, and had a lovely jaunt to Lundy Isle with my uncle. My grandmother also came down to stay (in one of the hotels) and altogether it was a brilliant time.

One of the highlights was when my parents decided to advertise the show by driving the caravan, towed by their jeep, whilst mother worked Muffin – Jack Whitehead had to hold onto her ankles to stop her falling off the stage as the caravan swayed around. They only tried that once!

About three years ago I gave a talk in one of the nearby villages in North Devon and a woman told me she remembered that summer very well. The puppets had been the highlight for her and she said that by the end of the month she was word perfect!

Muffin the Mule

Memorabilia Guide

Compiled by
Adrienne Hasler

Memorabilia Rarity Guide

Having devised the rarity guide I confirmed a fact that I suspect all collectors of Muffin the Mule items already know, very few items are particularly common. There are several reasons for this. Muffin the Mule gave the manufacturers of post-war Britain their first television-related marketing opportunity. Up until the mid-1950s many commodities were still in short supply and this resulted in low production numbers of many of the toys and goods. In today's market toys will be produced in their hundreds of thousands but, as far as I am aware, even the Moko Muffin production was far less than 100,000. Another problem involved packaging. Paper and cardboard resources were also in relatively short supply, particularly in the early years of Muffin merchandising, and companies like Chad Valley adapted whatever material was available to package the items. The teaset can be found in boxes of plain colour or patterned, each having the teaset label applied. Early post-war austerity also influenced the artwork on packaging and that is reflected in the relative desirability of objects.

The impact of the war is also reflected in the fact that many toys were thrown away or given to other children. Parents whose lives had been influenced by the trauma of the war were less likely to cherish mere objects. I have personally been contacted by quite a few people who wish to replace toys they owned as children. Their parents had disposed of the toys unable to comprehend the emotional attachment. Emotional attachment has probably created the demise of the item that I would most wish to find….a pair of Muffin slippers. One owner wrote to tell me of his slippers. Given to him on his stay in hospital, the slippers were his pride and joy. Having grown too big for them the slippers became puppets and were so well loved that they eventually fell apart. I suspect that if the majority of slippers did not suffer the same fate then they are likely to have fallen victim to moths.

With reference to values - the advent of the internet has given collectors greater access to auctions and dealers. This has resulted in a much more volatile market with the consequence of great variation in the price of similar items. Values are of course affected by many factors. These are commonly: condition, whether the item has its original box or wrapper, production numbers of item, whether the item crosses collecting boundaries, durability and desirability.

In any field of collecting condition affects values. Collecting Muffin the Mule items conforms to this trend. Value is further enhanced by original packaging and in some cases this may double the value of the item relative to an unboxed version.

For example the 'Moko Muffin Junior' was made in large numbers and children have played with the majority of these toys. Play-worn examples are readily available but are still popular. They may be bought to replace the toy lost from childhood or collected as a good example of the Muffin image. Boxed examples in mint condition are highly desired by collectors of die-cast toys, in particular those who collect Lesney/ Moko products.

Fragile items may have been manufactured in huge numbers, for example balloons or transfers, but by the nature of the materials used it is unlikely that many (if any) still survive. If survivors do exist they may not cost as much as another item which is sought after because, for example, it has attractive artwork. Of course, for almost every collector, survivors are, by definition, the few remaining items and are therefore deemed even more desirable to the dedicated collector.

Does it fit in with the rest of the collection? A decorative and easy-to-display item will be a sought-after item. If it is too large to be accommodated by the majority of collectors it will be less expensive even if rarer than a small and attractive piece. The amount of artwork on the packaging also affects desirability. Some items, for example, the Pin the Tail game has very basic graphics and colour. This item commands a lower price than the Bagatelle which has a highly decorative outer box. Desirability can be influenced by the crossing over of collecting fields. Again, the Moko Muffin Junior is a good example. This toy is collected by international collectors of Lesney/Moko items thus creating a vast market of collectors.

Other items, for example most of the theatre programmes, have a more limited market. However, the programme which lists Audrey Hepburn's first stage appearance is highly sought-after and is therefore expensive.

So, with all of these factors in mind, I have created a rarity guide and a relative value scale. Please note that this is a guide. Prices will vary and items thought to be rare may come onto the market in greater numbers.

Rarity Grade

★★★★★	Extremely rare, if any have survived.
★★★★	Rare in number and difficult to find.
★★★	Although made in large numbers, not easy to find.
★★	Fairly easy to find.
★	Readily obtainable.

Relative Value Grade

£££££	Sells at a high price
££££	Sells at a medium – high price.
£££	Sells at a medium price.
££	Sells at a low – medium price.
£	Sells at a low price.

1. Advertising *** £££
- Annette & Muffin were the 'faces' of Bush television and they appeared on two sizes of screen cover, 9" and 13". The covers could be fitted onto the front of the televisions displayed in electrical shops thus drawing the child's attention to that particular make of TV.
- This could be considered to be the first use of 'pester' power using children to influence parental decisions!
- Annette and Muffin (with Ann) made visits to the Bush factory and met with the production workers.
See *TV Comic no21 (28th March 1952)*.

2. Apron *** £££
- This was created to commemorate the Coronation of Queen Elizabeth ll. The apron decoration tells the story of 'Muffin and Co's' visit to the Coronation.
- Manufactured by JWS Hislop: - a 'Kumfy Kiddie Wear' product.
- The apron was available in both plastic and taffeta.
Advertised in *TV Comic no 68 (20th Feb 1953)*

3. Badges
First mentioned in *TV Comic number 4 (30th Nov 1951)*, the first Muffin Club 'shoe' appeared in issue *TV Comic number 5 (7th Dec 1951)*.
- 5 Muffin 'shoes' were required plus 1/- to become a Muffin Club member.

- Three variants of the *Club badge:* * £
There appears to be a 'boy' version with the normal saddle & a 'girl' version with a 'corset' type of saddle!

The *Road Safety Award Badge* ** £
- This badge required an application of 1/3d and appeared for the first time in *TV Comic 56 (28th Nov 1952)*. Because of her own serious road accident, Annette Mills was extremely forceful in her concerns over child road safety & children were awarded this badge. It could be removed by parents if they felt that the child was not safety conscious enough!

4. Bagatelle **** £££££
- Manufactured/marketed by Chad Valley c1955
- The game is created in the base of the box. Box size: 19" x 8½" (48.3 x 21.5cm).
  Post war packaging of toys still had to be carefully designed to be as efficient as possible, with no wastage.
- Muffin Syndicate 1949

5. Balloons ***** ££
Mentioned in *John Bull (9th September 1950)*.

6. Books
Author: Annette Mills
with dust jacket where originally present ** ££
without: * £

Illustrated by: Molly Blake. Published by: University of London Press
Muffin The Mule 1949
More About Muffin 1950
Muffin and The Magic Hat 1951
Here Comes Muffin 1952
Muffin At the Seaside 1953
Muffin's Splendid Adventure 1954

My Annette Mills Gift Book Undated
Illustrated by: Molly Blake, George Fry & Edward Andrewes
Published by: Heirloom Library, London * £

Ann Hogarth
The Red Muffin Book 1950 Illustrated by: Stanley Maile ** £
The Blue Muffin Book 1951 Illustrated by: Molly Blake & Neville Main ** £
The Green Muffin Book 1952 Illustrated by: Molly Blake, Neville Main, Jenetta Vise & Stanley Maile ** £
The Purple Muffin Book 1954 Illustrated as above, plus Comerford Watson ** £

Published by: Hodder & Stoughton Ltd / University of London Press
Meet Muffin The Mule 1954 Illustrated by: Neville Main. Published by: University of London Press ** £

Ann Hogarth/Annette Mills Merry Muffin books
** £ 1954 Illustrated by: Neville Main. Published by: Brockhampton Press
Muffin and Peregrine
Muffin's Birthday
Muffin and Louise
Muffin's Thinking Cap (AH only)
Muffin Climbs High (AH only)
Muffin Sings A Song (AH only)

1955 - **De Luxe Hard-Back** version of the Merry Muffin books. Sold for 4/6d *** £££
It appears from recent information recorded on the internet that these books were used in Australian primary schools as books for young readers.

Muffin's ABC Book *** ££
Available in hard back and soft cover.
Collins Wonder Colour Books.

Written &/or Illustrated by Molly Blake * £ 1958
Muffin's Own Story Book Publisher: Collins

Neville Main: Author & illustrator * £ 1952
Publisher: Brockhampton Press
Muffin And His Friends
Muffin On Holiday
Muffin Makes Magic
Undated *'Muffin The Mule'* illustrated in *"Champion Story Book"* by Neville Main * £ Publisher: Collins

Muffin also made appearances in
Film & Television Parade (Undated) * £
Television Annuals 1954/56 * £
BBC Children's Hour Annual 1952 * £
Undated *Collins Children's Annual* x 2

7. Bracelet charm *** ££
- A tiny 'charm' for attachment to a bracelet.
- The neck, head and legs are articulated and there is bridle and saddle detail to the figure.
- Maker unknown.

8. Brooch
- **The plastic brooch.** *** £
- Sold by the Hogarth Puppet Theatre at their touring performances alongside the Pelham Puppets. See reference in *Puppet's Progress* by Jan Bussell.
  Height: 1½" (3.8cm)
- **The enamel brooch.** ** ££
- Sold through Thomas Horton Ltd., Birmingham.
- Advertised in *Television Weekly (1st Dec 1950)*.
- Height: 1" (2.5cm)
- Original price 3/6d
- Licensed as featured by Miss Annette Mills and Miss Ann Hogarth.

9. Busy Box **** £££££
- Manufactured/marketed by Chad Valley c1953.
- Includes sewing & colouring items plus Muffin skittles.
- Chad Valley had its own printing works for box covers and labels.
- The company utilized various toy components in several of its products. For example the Muffin skittles were sold boxed for very young children but packed in the Busy Box with other items.
- Muffin Syndicate 1949

10. Buttons ** £
- Created in a variety of colours and two sizes, the buttons were hand painted.
- Sizes: ½" diameter & 0.7" diameter (1.25cm/1.8cm)
- Manufactured by Jason Buttons, London SE24

11. Calendar ***** ££

Mentioned in the *Evening Standard* in an article about **'Kiki'** the Kiwi. October 1952.

It appears that the Kiwi was originally named as KiKi in The *Evening Standard* article from October 1952. He was Kirri by 19th December 1953 when he is photographed with the Hogarth Puppet line-up for the *'Illustrated'* publication. In *Puppet's Progress* (1953) Jan Bussell also calls him Kirri.
Just to make matters even more confusing, the Huntley & Palmer's tin names him as Kerri!
Derek McNally confirms that Kirri was the correct name for this puppet.

12. Cards

- Birthday ** £, Christmas *** £ and
  Party invitations *** £
- Three styles of card.
- Only the cards with the gilt age portion are printed with the publisher's name.
- The similarity of the artwork suggests that all cards may possibly be by the same publisher.
- James Valentine Publishing Company, Dundee. Undated.
- Muffin Syndicate 1949

13. Carousel and Coin Rides *** £££

- Created by Edwin Hall Ltd. 1953
- The company held sole rights to "showman's equipment" in respect of Muffin and his fellow animals.
- Initially the rides were constructed of wood and were fixed on carousels or various other types of fairground ride. They were also equipped with a motor as a coin ride. Later versions of the coin ride were made in fibreglass.

14. Ceramic model **** ££

- Sitting model of Muffin.
- Unmarked except for a hoof mark underneath.
- Originally assumed to have been especially made in small numbers as a gift to Annette Mills and Ann Hogarth. However, *The Evening Standard (Oct 1952)* mentions ceramic Muffin figures.

15. Christmas Crackers ***** ££££
- Produced by Caleys and advertised in the *TV Comic no.160 (27th November 1954)*
- Contained tiny illustrated books 1¼" x 2" (3.2cm x 5.1cm), the book was entitled 'The Smallest Muffin book in the World'.
- The books contain different stories including: 'Muffin's Cure for a Cold', 'Muffin and the Sea Serpent', 'Muffin foils the Smugglers' and 'Muffin the Detective'.
- These tiny books carry illustrations of single characters on one page and the colour cartoon story on the other.
- Muffin Syndicate 1949.

16. Christmas Tree Lights *** £££
- Unmarked although possibly manufactured by Pifco.
- Undated although an advert for Pifco Christmas lights in *'The Wireless & Electrical Trader'* (30th September 1950) shows a similar style.
- Featured: - Muffin, Louise, Grace and Peregrine.

17. Comics & annuals * £
- Muffin featured prominently in *TV Comic* throughout the 1950s and remained as a strip cartoon until 1961.
- See *TV Comic Time Line* page for a brief description of important comic dates & information on those dates.
- The *TV Comic Annuals* 1952-1960 all contained Muffin stories. Published by Beaverbrook Newspapers Ltd.
- *TV Mini Books No 1, 6, 8, 12*. Undated but first advertised in *TV Comic no 100* – Mini books numbers 1-3.
- Illustrated by Neville Main
- Published by: The News of the World Ltd.

18. Cutlery
- Manufactured by Thomas Turner and Co c1953 (see also Napkin ring and Penknife).
- The spoon and fork have Muffin the Mule and musical notes on the tip of the handle. Length: 5.7" (14.5cm) *£
- The bone-handled knife has Muffin the Mule and musical notes etched into the blade. *** ££

19. Derby Race Game ** ££
- Manufactured by: Glevum (Roberts Bros. Gloucester, Ltd).
- This company was bought out by Chad Valley in 1954 so it would appear that the game pre-dates that year.
- Muffin Syndicate 1949

20. Die-cast figures
Manufacturers:
- Argosy – sold in a boxed set with Peregrine, Louise & Peter. 1953. **** ££££
  Rarest of the die-cast figures produced.
  2" (5cm) high. Muffin Syndicate

- 'Luntoy (London Toy Company) Children's Television' figures *** £££
  Made by Barrett & Sons 1952
  Considered less rare although box often missing.
  Sold in 'television' boxes, Peregrine the Penguin also available.
  2" (5cm) high

- Sacul *** £££
  Muffin figure available 1951.
  Rare and usually tail-less.
  Believed to be boxed.
  2" (5cm) high

- Wend-al *** £££
  Two sizes available. The larger version having hinged legs.
  Made in Dorset and sold unboxed.
  Cast in aluminium.
  1.8" (4.6cm) & 3.5" (8.9cm) high

21. Drawing Slate **** ££££
- Chad Valley item 1955.
- Utilized the same illustrated sheets as the Picture Cubes.
- Story book size 6"x 6" (15.2cm x 15.2cm)
- Advertised in *W. Fielden and Sons Ltd* catalogue in 1955. Retailed at 2/11d
- Muffin Syndicate 1949

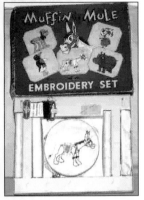

22. Embroidery Set **** ££££
- Manufactured/marketed by Chad Valley.
- Five sewing kits to embroider: Muffin, Louise, Peter Hubert & Peregrine.
- With a message in the lid from Muffin as to how to create a handkerchief sachet or use the finished items as dressing table mats.
- Muffin Syndicate 1949

23. Ephemera
- Various signed photographs of Annette and Muffin were sent out to children on request and book signings took place around the country. In *TV Comic no 68 (20th Feb 1953)* it was announced that Muffin and Annette would visit the Ideal Home Exhibition on 18th March 1953 to sign books. * £

- Theatre flyers and programmes.
  The most sought-after programme is for Cecil Landau's Christmas Party because it not only introduces Muffin the Mule but Audrey Hepburn too! **** £££££
- Muffin also featured in menus relating to Christmas meals at the Cumberland Hotel in 1951 & 1952. *** £££
- Theatre venues included Lyric Theatre, Hammersmith; Theatre Royal, Bath; Institut Français (23rd Oct 1950 for a special performance for the Puppet Guild members); Vaudeville Theatre; Wimbledon Theatre.
- In addition, *TV Comic no 82 (29th May 1953)* records Muffin as appearing at The Hogarth Theatre Tent, Waterlow Park, Highgate & then at The Rookery, Streatham Common in June.
- Muffin films were recorded as showing: Sheffield News Theatre; Eros New Theatre, Piccadilly; Waterloo News Theatre, London *TV Comic no 65 (30th Jan 1953)* and in Sheffield, Piccadilly, Dalston, Margate, Croydon, Southampton, Tooting & Hammersmith.

24. Fabric **** £££
- Maker - unknown
- Cotton fabric available in at least two colour-ways: pink and blue.
  The fabric shows many of Muffin's friends and Muffin.

25. Glove Puppet *** ££
- Made by Chiltern Toys.
- Puppet has a fabric label with red printed 'Chiltern Hygienic Toys'.

(Fabric labels were introduced in the 1940s - the red printed label often sewn in the side seam, and then in the 1950s a white label, with blue print, was used. Information from web page. See bibliography).

This puppet therefore probably dates to 1948/49.

26. Gramophone **** £££££
- The Nursery gramophone was manufactured by the National Band Gramophone Company (London) according to the label.
- Constructed with a wooden carcase, the wind-up gramophone is covered with a 'Rexine' cloth and decorated on the top, front and inside with transfers of Muffin and his friends.
- Muffin records were manufactured in 1949 and in 1950, therefore it would seem likely that the gramophone was produced at this time.

27. Hair Ribbon **** £££
- Advertised in *TV Comic no 9 (27th June 1952)*
- Manufactured by: S.Levy & Sons Ltd, Harrow Weald.
- Two versions available. One - multi-coloured features Muffin plus friends, the other features Muffin against blue clouds kicking up his heels.

28. Handkerchiefs ** ££
- 1950s.
- Boxed hankies. The Television Novelty boxed set originally retailed at 4/11d advertised in *Robinson & Cleaver Christmas Gifts catalogue* (undated).
- Another version has a box decorated with drawings of Muffin & his friends.
- Muffin Syndicate 1949

29. Hobby horse ***** ££
Listed in *Classic Toys Magazine (1995)* article by P Talbot.

30. Jigsaw Puzzles *** £££
- A 'Moko Toy' set of Puzzles.
- Designed by Molly Blake. Manufacturer & date unknown.
- Muffin Syndicate

31. Lamp *** ££
- Two versions of the plastic/bakelite version are known. One is on a short base and one on a taller version.
- Height of short lamp is 25.5 cm (10"). The lamp has transfer decoration of Muffin and three friends.
- Also, a lamp with a fabric shade with characters printed on it. **** £££££
- Lamp bases (and plastic shade) possibly manufactured by Pifco c1951 (see Christmas lights)

32. Ludo Game ** £
- Manufactured/marketed by Chad Valley

33. Modelling kit **** ££££
- Manufactured: under licence to Burt and Edwards (Hull) Ltd by LF Booth & Co Ltd., Hull makers of 'Castime' products.
- Original cost 15/9d. Extra moulds 2/5d.
- Box contained moulds for Muffin the Mule, Wally the Gog and Peregrine the Penguin.
- Other moulds available: Peter the Pup, Monty the Monkey, Zebbie the Zebra, Hubert the Hippo and Oswald the Ostrich.
- Muffin mould height: 3½" (8.9cm) Peregrine mould: 4" (10.2cm)
- Date c1954
- Sold 'by permission of Annette Mills and Ann Hogarth'.

34. Money box *** £££
- Chad Valley item c1950 tin plate lithographed. Height: 3" (7.6cm)
- Muffin Syndicate 1949

35. Napkin Ring *** ££

- Manufactured by cutlers Thomas Turner and Co.
- Muffin design is engraved as on the knife blade and penknife.

36. Paint Box ** ££

- Produced by: George Rowney & Co. (Artists' Colourmen & Pencil Makers)
- Muffin Syndicate 1949

37. Painting Book ** £

- The Whopper Muffin Painting book published by The Children's Press, London and Glasgow c1956
- 64 pages to colour

38. Patterns *** ££

- Sewing pattern advertised *Television Weekly (1st Dec. 1950)* and *TV Comic No 16 (22nd Feb 1952)*
- Price: 1/2½d

- Weldon's knitting pattern
- Undated

39. Pencils *** ££

- Produced by: George Rowney & Co. (Artists' Colourmen & Pencil Makers)
- Box of 10 pencils
- Each of the pencils has Muffin and his friends as decoration. Box decorated with characters.
- Muffin Syndicate 1949

40. Penknife *** ££

- Single bladed penknife manufactured by Thomas Turner and Co.
- Size of handle: 3" (7.6cm) long
- Muffin decoration engraved on both sides of the handle.

41. Photographers models **** £££
- Recorded by Annette Mills in *TV Comic no 55 (21st Nov. 1952)*, captioning a photograph of a young girl on a giant model of Muffin, she tells readers that such giant versions would be found at the seaside next summer.
- 1960s photograph of young boy taken by Mr G W Felix at Bognor Regis. I have been told that the same Muffin actually appears in the 1962 Tony Hancock film The Punch and Judy Man filmed in Bognor Regis. The boy (Mr Felix's son, Geoff) is now a professional puppeteer including Punch & Judy, published author and, most importantly, a member of the Muffin the Mule Collectors' Club.

42. Picture Cubes **** £££
- Chad Valley product
- Uses the same pictures as the Tracing slate
- c1952
- Muffin Syndicate 1949

43. Pin the Tail on Muffin ** ££
- Produced by Chad Valley: Card with Muffin and a cotton tail
- Undated
- Muffin Syndicate 1949

44. Playing Cards * £
- Pack of 44 Playing Cards
- Published by Pepys (Castell) Cards 1955
- Muffin syndicate 1949

45. Postcards ** £
- Set of six cards. Numbers 5379 - 5384 inclusive.
- Published by J Salmon & Co, Sevenoaks, Kent c1956
- The cards were part of a series including Andy Pandy, Flowerpot Men, Woodentops and Sooty.
- Each of the themed sets consisting of six cards.
- Muffin illustrations by Neville Main
- Muffin the Mule titles: Muffin At The Seaside; Muffin's

Caravan Journey; Muffin Playing Cricket; Muffin's Picnic; Muffin Gets A Wetting & Muffin Fishing.
- Muffin Syndicate

46. Pouffes **** £££
- There are two designs and two sizes -13" (33cm) & 14.2" (36cm) in diameter.
- Leatherette decorated outer cover and straw filled.
- No makers label.
- Designs – 'Muffin the Mule' and 'Muffin's Friends'.

47. Projectors and films ** ££
- TV Cine and Mini Cine projectors and film strips.
- Manufactured by Martin Lucas Ltd.
- Advertised: *TV Comic no 53 (7th Nov 1952)*

48. Puppets
- Zinc metal 'Moko Muffin Junior' puppet designed by Alfred Gilson in 1950. * £ unboxed ** ££ boxed
- 6" high with four finger rings to work him.
- Manufactured by A. Gilson Ltd and marketed by Richard Kohnstam (Moko).
- Following A Gilson's emigration to the US, Kohnstam approached Lesney, who had rented accommodation from Gilson, to manufacture the toy 1951-1955.
- It has been estimated that 60,000 puppets were sold at the time.
- Although sold in huge numbers, these puppets are highly sought-after today.
- Probably the best representation of the illustrated version of the television character made.
Marked Muffin Syndicate Ltd.

- Plastic Moko 'Peregrine' puppet Size: 5" (12.7cm) high. *** £££££
- Much rarer puppet.
- Peregrine is able to flap his wings, move his feet and beak.
- By permission of Ann Hogarth and Annette Mills.

- Pelham puppet ** ££ unboxed. *** £££ boxed in the Muffin the Mule box.
- Bob Pelham was a close friend of Jan Bussell and in 1947 Jan advised him on a control and stringing design to allow amateur puppeteers to work their marionettes (stringed puppets).

- In his book, *Puppet's Progress* Jan Bussell speaks of selling Pelham puppets at his shows. It seems that, initially, these did not include Muffin the Mule puppets.

- In 1957, Pelham's created a special Muffin the Mule box for the Pelham Muffin. Later, Muffin puppets were sold in the standard Pelham box.

- Muffin the Mule puppets were sold between 1957 & 1972.
- Originally, the puppets were sold from Bob Pelham's shop premises but a successful demonstration in Hamleys toy store in Regent Street, London led to the puppets being demonstrated and sold in Gammages and Harrods.
- Advertised in *Harrods 1957 Christmas catalogue*
- The 4" (10.2cm) high Pelham Muffin shares a body with the Little White Bull!

- Jan Bussell wrote a book of 12 plays for Pelham puppets including one for Muffin.
- Muffin Syndicate Ltd.

- In 1954, Muffin appeared on the back of Kellogg's Rice Krispies packets encouraging children to 'Make your own puppets!'

49. Racing card/toy *** ££
- Maker and date: Unknown.
- Muffin moves as the race card is moved. He gallops to the winning post.
- It is advertised as one of a series including "Muffin's friends".
- Muffin Syndicate

50. Records *** ££
- 12" Decca double album no 1 1949 – DR 12770/1 – shellac
- 12" London double album no 2 1950 – DR13568/71 – vinyl
- 7" Oriole singles - shellac. Set of three - Muffin the Mule; Wash day & Monty's Swing Song.

51. Rugs *** ££

- Manufactured by: The Trail Trading Co Ltd., London, EC2
- "Muffin Land" rugs were sold in assorted designs and in blue, pink, green and beige.
- The designs are found in reverse giving at least two variations of each rug.
- They featured Muffin the Mule, Louise the Lamb and Peregrine the Penguin.
- After Zebbie the Zebra became part of Muffin's group of friends in December 1953, a rug depicting Zebbie and Muffin having a picnic was created.
- Original cost: 25/- Rug size: 36" x 21" (91.5cm x 53.5cm)
- Advertised in *TV Comic no56 (28th Nov 1952)* and in the *Vaudeville Theatre programme (Dec 1952)*, advertising "the latest rug from 'Muffin-land' featuring Peregrine the Penguin".

95

52. Scraps *** £££
- Printed in sheets for children to put into Scrap books, the scraps feature: Grace the Giraffe, Katy the Kangaroo, Louise the Lamb, Monty the Monkey, Morris & Doris the Field-mice, Oswald the Ostrich (plus Willie the Worm), Otto the Octopus, Muffin the Mule, Peregrine the Penguin, Peter the Pup, Polly the Parrot, Sally the Seal, Sam the Scarecrow and Wally & Molly the Gogs.
- This set probably gave children the most comprehensive set of Muffin the Mule characters of all of the merchandising available with the exception of the playing cards.
- Possibly printed by - AJ Donaldson (Publishers), Glasgow.

53. Sherry glasses
- Gifts to Ann Hogarth and Annette Mills.

54. Skittles **** £££££
- Chad Valley skittles 'for very young children'.
- The game consists of five cardboard skittles on wooden supports with three wooden 'bowling' balls.
- Muffin skittle is 5½" high (14cm)
- Other skittles are Louise the Lamb, Peter the Pup, Peregrine the Penguin and Hubert the Hippo.
- Muffin Syndicate
- Undated.

55. Slippers ***** £££££
- Mentioned in the programme for Cecil Landau's Christmas Party 1949 and by Jan Bussell in *'The Puppets and I'* published in 1950.
- Advertised in *TV Weekly (1st December 1950)* and in *TV Comic no 52 (31st Oct. 1952)*
- Available in: fawn, red or blue felt.
- Original price: from 9/3d to 11/3d depending on size.

56. Soaps
Made by - Cullingford of Chelsea.
- **Individual figural soaps** were made both in white or colour. ** ££
- The figures used were: Muffin the Mule, Zebbie the Zebra, Peregrine the Penguin, Louise the Lamb and Monty the Monkey.
- Advert appears in *'Enid Blyton's Magazine'* no 26, (Vol. 2) 22nd Dec. 1954.
- Plain soaps from 9d. Coloured versions 2s 3d.
- Sizes: Muffin, Louise and Zebbie 3½" (8.9cm) Peregrine 4" (10.2cm)
- **Soap tablets** – available in boxes of three or, individually. *** ££
- Tablets moulded with the figures of Muffin the Mule, Louise the Lamb or Peregrine the Penguin.
- Cost 1/2½d in the plain version or hand-painted at 2/1½ d each.
- Licensing: Muffin Television Puppets
- **Free Muffin transfers** could be acquired by sending one empty Muffin soap box and a stamped addressed envelope to the magazine. The offer was available in *'Enid Blyton's Magazine'* no 6, (Vol. 2) (17th March 1954).
See later number 62.

57. Song book * £

- Although billed as the First Muffin Song Book in fact it was the only one.
- Music written by Annette Mills. lllustrated by 'Bud' (Molly Blake).
- Publisher: Chappell & Co Ltd.
- Original price 3/-
- Undated although mentioned in *Bandwagon (Sept.1949)*.
- By permission of the Hogarth Puppets

58. Tablecloth *** ££

- The square cloth is unmarked and undated.
- Printed decoration has Muffin and friends at Muffin's Birthday and Muffin's School plus individual characters.
- Measurements: 48" x 48" (122cm x122cm).

59. Tankard **** ££

- 2.4" (6 cm) high pewter tankard etched with the same Muffin the Mule design as the Thomas Turner plated items.
- Bears 'touch marks' – the pewter equivalent to the silver hall marks although these marks have no legal standing.

60. Tea set and tea ware
- **Chad Valley Set: ** £££**
- Marketed by Chad Valley, the china tea set was manufactured by Wade Heath and Co.
- Wade, Heath & Co produced Art Deco nursery ware. The set uses the same style teapot as the Mickey Mouse teaset (dated c1936) and the shape remained in usage for nursery ware into the 1950s with the relevant transfers for sets for Muffin the Mule, Noddy and Prudence Kitten. The Muffin the Mule set utilizes the Deco shaped tea pot with a later style cup shape.
- Although the Disney items were marked with the backstamp 'Wadeheath England by permission Walt Disney', the Muffin tea set is marked only with the word 'England'.

- **Barretts of Staffordshire Teaware. **** ££**
- The company made a series of character sets consisting of a tea cup and saucer, plate and bowl. The Muffin image is more stylized than the normal one and the items were associated with a monkey, giraffe and deer. These are not believed to be part of the group of Muffin's 'friends'.

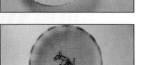

- **Opaque Glass Items ** ££**
- Plate, bowl/dish plus two versions of mug.
- Unmarked

61. Television set * ££

- Beeju Toy television set "complete with lighted screen".
- Manufactured by EVB Model Aircraft Ltd
- Original price 12/6d Extra films 1/-
- Advertised in *Television Weekly (1st Dec 1950)* and in *TV Comic no 17.(27th Feb 1952)*
- Copyright: Muffin Syndicate 1948

62. Tins

- The biscuits were manufactured by Huntley & Palmer whilst the tins themselves were created by Huntley, Boorne and Stevens (see Huntley & Palmer's web site). The tin manufacturers were a company which included as a founder member another member of the Huntley family.

- The tins & biscuits were sold in 1954.
- Characters illustrated: Muffin the Mule ** £, Peregrine the Penguin ** £, Louise the Lamb ** £, Monty the Monkey *** ££, Grace the Giraffe *** ££, Hubert the Hippo *** ££, Peter the Pup ** £, Zebbie the Zebra *** ££, Katie the Kangaroo *** ££, Oswald the Ostrich ** £ and Kerri the Kiwi *** ££.
The correct name of the Kiwi is actually Kirri.
- Copyright: Muffin Syndicate 1949

63. Transfers
Two types of transfers exist:
- **Free Muffin transfers** could be acquired by sending one empty Muffin soap box and a stamped addressed envelope to the magazine. These were small transfers to enliven the tablet soaps. ***** £££**
- The offer was available in *'Enid Blyton's Magazine'* no 6, (Vol. 2) (17th March 1954).

- **The Calco Company Ltd.** produced simple transfers that could be used as a 'means of educating children in the simple art of decoration'.
  The transfers could be used to decorate their furniture or personal belongings. ***** £££**
- Copyright: Muffin Syndicate

64. Wall paper ***** £££**
Produced by Crown.

65. Wooden toys

Luntoy Pull-along toy Marked Luntoy Supertoys. Made in England. **** £££
- 10½"(26cm) long x 8½ (19cm) high.
- No copyright information.

Wooden articulated Muffin the Mule.
- 5½" (14cm) to shoulder. **** ££££
- Illustrated in *The Sunday Graphic (27th Nov 1949)* in an article looking at toys for Christmas.
- Retailed by Selfridges for 19/6d..
- Unknown maker.
- In *TV Comic no 29 (23rd May 1952)* a young boy is shown holding one of these toys in his lap.
- No copyright information.

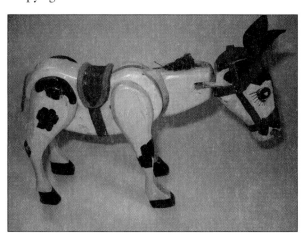

This is Not Muffin!

There are a few items that are often described as being Muffin but are not. I have tried to provide full and accurate information on all items and I must apologize to those adults who have cared for (and loved) items that I am now going to identify as 'not Muffin'.

The main one is the mule with a 'turnip ring' to make him open his mouth and turn his head.

Made in several colours and on two different types of base, this little mule is actually 'Moody the Mule'. Manufactured by Gantoy Ltd.

I have established that Triang (Lines Brothers) did not manufacture any toys under licence to The Muffin Syndicate, also that the little wooden pull-a-long toy by Chad Valley was not, in fact, Muffin the Mule.

A wooden pull-a-long toy was produced by Luntoy as shown in the guide.

With the sale of the 'Calco Muffin Transfers', it was possible to add transfers to items not sold under licence.

I have seen one such example in the form of a tall pottery jug bearing the Father Christmas Muffin transfer and others may well exist.

Chad Valley toy.

Original Calco Company Ltd transfer seen on jug.

Modern Muffin Collectables

It seems that Muffin the Mule has remained in the collective mind of the population.

The BBC kept up their close association with Muffin by inviting Muffin and Ann to the closing of Lime Grove. When stamps were being designed to commemorate fifty years of children's television broadcasting naturally Muffin and Annette were included. The set of stamps was titled 'BIG STARS FROM THE SMALL SCREEN'. It does appear that this set of stamps had its critics. Questions were asked in the House of Lords (see Lords Hansard Written Answers Wednesday, 15th May 1996) about the money spent on copyright issues regarding Muffin although there do not seem to have been questions asked about any of the other icons of children's television. I understand that there were people who wrote letters to the Press expressing their negative views.

- **Commemorative Stamps**
  Muffin and Annette from a BBC photograph taken in 1952. The stamp was designed by Tutssels and issued by the Royal Mail on 3rd September 1996. The stamp design was also used for a 'Benham Silk' limited edition presentation envelope.

- **Scope Envelope.**
  Also issued as a First Day Cover on 3rd September 1996.
  No information about the designer.

- **BBC Notepad**
  Licensed by BBC Worldwide.
  The note pad has the Muffin the Mule stamp as its cover using the same format as the Post Office 'Stamp (post) Card Series'.

- **Tea towel** from the Museum of Moving Image which opened in 1988 and closed in 1999.
  The tea towel has no information other than it is UK made.
  Other images include: Andy Pandy, King Kong, Daleks, a Spaceman on the Moon, Laurel and Hardy, Dorothy and the Scarecrow from the Wizard of Oz and triumphant English footballers.

- **Danbury Mint Plate**
  One of twelve plates issued in the 'Golden Age of Children's Television' series.
  Measures 8" (20cm) in diameter.
  Decorated with a 22-carat gold band around the rim of the plate.

Muffin the Mule Collectors' Club items:
Exclusive to Club members.

- **Membership badge**
  This image of Muffin appeared on letter headings used by Annette Mills and Ann Hogarth. Sally McNally also used the same design and in 1999 granted permission for the Club to use it on headed paper, the newsletters and to create the design for the badge.
  The badge is manufactured by M & L Promotional Products Ltd., 5 Queen Street, Mirfield, West Yorkshire. WF14 8AH and each badge is numbered and relates to the individual's Club membership card.

- **Sweat shirts/polo shirts**
  Using an embroidered image in the Neville Main style of drawing.
  Originally offered in 2001 shirts will again be available from 2005.
  Produced by Actifwear, 1 Cross Street, Market Harborough, Leicestershire. LE16 9ES

MODERN MUFFIN COLLECTABLES

- ### Christmas Cards
  Designed and created by Paul Robbens. Photography by Adrian Wroth.
  Each card has been created using models.

- ### Jumping Jack Muffin
  Designed and created by Paul Robbens of 'PRSFX'.
  5" (12cm) long and 4½" (11cm) high. Hand-made from Russian Birch. Hand-painted.
  Cost £28.00

- ### 'Nodder' Muffin
  Designed and created by Paul Robbens of 'PRSFX'.
  25 were made for Club members.
  Maverick Entertainment Group plc requested another 6.
  Cast in polyurethane resin. Hand-painted and finished with hand-made bridle and saddle.
  4½" (11cm) long and 6½" (16cm) high.
  Cost £45

- ### Muffin puppet
  Designed and created by Paul Robbens of 'PRSFX'.
  Cast in polyurethane resin. Hand-painted and finished with hand-made bridle and saddle.
  Strung in the same way as Muffin himself with a split control bar.
  8" (20cm) long and 7½" (19cm) high
  Cost: £120

- **Muffin Artwork**
    Illustrator Walt Howarth.
    Produced in 2002. Commissioned by Steve Penny with permission from Sally McNally.
    Three sets of prints produced. Purchased unframed or mounted and framed.
    Prices are for unframed prints.

Set 1 - on normal matt paper A4 sized paper.   Edition size: 20 Price: £12
Set 2 - A4-sized printed on thick glossy paper. Edition size: 20 Price: £12
Set 3 - A3-sized prints on the glossy paper.   Edition size: 18 Price: £45
Available through: Pure Nostalgia, 121 Chadwell Road, Grays, Essex RM17 5 TG

- **New animated version of Muffin the Mule 2005.**

- **60th Anniversary Annual 2006**
Available September 2005 and acquired through Maverick Entertainment Group plc.
Combines the new animation style of Muffin with some information about vintage Muffin and original merchandising.
Cost £6.99

- **4 hardback storybooks**
Available September 2005 from Maverick Entertainment group Plc.
"There's No Place Like Home"
"Who-dunnit...?"
"Wish Upon A Star"
"Muffin's Mules United"
 Cost £2.99

Muffin Syndicate

I am indebted to Derek McNally for the following information about the company set up by Ann Hogarth and Annette Mills to allow companies to license Muffin the Mule products.

The Muffin Syndicate (Ltd) came about after various manufacturers approached Ann Hogarth and Annette Mills offering to produce Muffin the Mule products if the owners (The Hogarth Puppet Theatre as owners of the puppets & Annette Mills as owner of the songs) would pay for the privilege!

Ann and Annette soon realized that this was not how things should work and it was not long before a variety of manufacturers were lining up to license toys and other items.

The earliest toy box with copyright recorded is Muffin Syndicate 1948. This appears on The 'Beeju' television which was advertised for sale in TV Comics dated 1952.

The Moko Muffin Junior puppet was marketed in Selfridges in July 1950 and bears the mark Muffin Syndicate Ltd. Items may bear either Muffin Syndicate Ltd or Muffin Syndicate and are dated for either 1948 or 1949 or have no copyright information whatsoever.

It seems that following Annette Mills' death, Jan Bussell became the second half of the Syndicate until the Bussells found it too tiresome to continue and closed the Syndicate for marketing purposes in the early 60s. As the Hogarth Puppets continued to tour and travel internationally, Muffin was absorbed back within the Theatre until Ann Hogarth's death.

It seems that, following Jan Bussell's death, Ann Hogarth destroyed the papers relating to the Muffin Syndicate. Following Ann's death, Sally, her daughter, set up her own company to protect Muffin's copyright.

In 2003 a licence was granted for a limited period to Maverick Entertainment Group Ltd and they have developed a new style Muffin whilst retaining his essential characteristics. Muffin's friends have also been included within the new version.

With Sally McNally's death in 2004 Muffin's copyright has been inherited by her husband, Derek McNally.

Muffin the Mule

TV Comic Memorable Events

Date	Number	Event
1951		
9th November	1	First edition of the TV Comic
7th December	5	1st 'shoe' appears - to collect to join the Muffin Club. 5 required plus a fee of 1/-
13th December	6	Muffin 'coat of arms' on the Muffin Club membership card
1952		
4th January	9	Entry form to join the club as a 'Founder Member'.
22nd February	16	'Make your own Muffin' Style pattern advertised
29th February	17	'Beeju' TV advertised with 12 toys being offered as prizes in a handwriting competition. Annette Mills announces that 50,000 members enrolled in the Club within the first week.
		TV sets would be sent to individual homes as well as hospitals & children's homes.
28th March	21	Annette mentions two sets of twins enrolling as Muffets.
		Muffin off to Australia for 6 months and reassures children that films of Muffin will be shown on Sundays.
11th April	23	Scotland gets a TV transmitter.
2nd May	26	Club members have enrolled whilst living in Fiji, USA, New Zealand, Egypt, South Africa, Gibraltar & Germany.
16th May	28	Little (local) Muffin Clubs run by children are appearing.
		In Nottingham, the members of the 'Beeston Muffin Club' includes Muffet number 100,000 – Guy Spivey.

27th June	34	Hair ribbon advertised
25th July	38	Muffin & His Friends by Neville Main advertised at 1/6d
8th August	40	Muffet number 150,000 – Julie Rose Corbin is enrolled
5th September	44	Mini-cine projector advertised
19th September	46	Wales and the West Country get TV Transmitters
10th October	49	Muffin returns from Australia on SS Orantes. Annette reports on nine Muffets in one family – the children of Mr & Mrs Sigley of Leek, Staffordshire.
17th October	50	Muffin celebrates his birthday
7th November	53	TV Cine advertised
14th November	54	TV Comic Birthday Issue
21st November	55	Photograph of girl on a photographer's model Muffin with Annette noting that Muffins would be found at the seaside photographers next summer.
28th November	56	The Road Safety badge award is announced. Muffets are recorded in North Africa and Saudi Arabia.
12th December	58	Annette announces the Muffin Show at the Vaudeville Theatre running for four weeks from 23rd December. All Muffets would receive a gift.

1953

2nd January	61	Road Safety badge application form
30th January	65	Muffin films were showing at the Sheffield News Theatre; Eros New Theatre, Piccadilly; Waterloo News Theatre.
		The comic expands from 12 to 16 pages.
6th February	66	Annette Mills reports on her Daily Mail National Children's Award.
13th February	67	Annette tells the readers that the Club gains its

		youngest Muffet – a newborn baby in New York, USA and records another Muffet living on the Gold Coast of Africa.
20th February	68	Muffets model the Coronation apron. Muffin films are showing in Sheffield, London, Dalston, Croydon, Southampton, Tooting and Hammersmith.
		Children are encouraged to visit the Ideal Home Exhibition on 18th March to meet Muffin and Annette for a book-signing session.
17th April	76	Muffet number 200,000 is identified as Peter James of Leicester.
15th May	80	S.L.Ford of Hampshire is made a Honorary Senior Muffet when he donated the special (expensive) aerial required to install the television given to the Sunshine School, Alvestoke.
22nd May	81	The 50th television set is installed in Evelina Ward, Guys Hospital. It is announced that Muffin and the Hogarth Puppets are to appear at The Theatre, Kettering.
29th May	82	It is announced that Muffin and the Hogarth Puppets are to appear at 'The Hogarth Theatre Tent' in Waterlow Park, Highgate and then at The Rookery, Streatham Common.
25th September	99	Sooty joins TV Comic. The Purple Muffin book is advertised, following on from the Red Muffin Book (1950), The Blue Muffin Book (1951) and the Green Muffin Book (1952).
2nd October	100	Advertising for the four books above plus Annette Mills' set -
		Here Comes Muffin priced at 7/6d, the other books at 6/6d. The TV Comic Mini Books 1-3 are advertised at 6d each.

1954

18th September	150	Annette presents a television to the Crippled Children's Hospital, Plaistow and encourages readers to buy the new TV Comic Annual.
27th November	160	TV Comic Annual advertised at 7/6d. The

Muffin Club 'Road Safety badge' is advertised in colour, also an advertisement for 'Caleys Crackers' containing the 'Smallest Muffin book in the World'.

After Annette Mills' death, Ann Hogarth continued to write for the Club page. Sally and Derek McNally were subsequently enlisted and wrote stories for the Muffin pages. Neville Main continued to illustrate both Muffin the Mule and other stories.

1955

9th July	193	Sooty takes over from Muffin on the front cover.

1956

24th March	229	The 'Sooty Friendship Circle' welcomes Muffets.
31st March	230	The announcement that Muffin is back on television on ITA.

1960

12th March	430	The Muffin Club is still in existence! Muffin appears in the TV Comic Annual but it is obvious from the much sparser artwork that TV Comic is no longer a lavish publication.

1961

11th March	482	Muffin has his last story in the comic.

I am grateful to the following website www.oxforddiecast.co.uk for the information for 1961.

Photography Credits

The illustrations on the memorabilia pages appear by kind permission of the following:

    Caroline Carlisle: Zebbie Rug

    Paula Connelly: Coronation Apron

    Sandra Fallon: Fabric shade nursery lamp

    Adrienne Hasler: Memorabilia images

    Edward Hasler: Memorabilia images

    Items were lent for photographing by:

    Hazel Clements

    Jean Dudley

    Pete Jackson

    Patrick Talbot

    Adrienne Walsh

Tracing Copyright Owners

Every effort has been made to trace the copyright holders of old material. Where these efforts have not been successful, copyright owners are invited to contact the author so that their copyright can be acknowledged.

Bibliography

Bandwagon Sept 1949:
Norman Kirk Publications.
Grand Buildings, Trafalgar Square, London WC2
BBC Year Book 1946:
The Hollen Street Press, London W1
Classic Toys Magazine 1995:
MICA, Southend on Sea
Decca Records Book 1949:
Mears & Caldwell Ltd, Cramner St., London SW9
Enid Blyton Magazine Mar. & Aug. 1954:
Evans Brothers Ltd., Montague House, Russell Square London WC1
GB Film Library 1956/57:
1 Aintree Rd, Perivale, Greenford, Middlesex
'Gifts for Good Children. The history of Children's China part l l. 1890-1990':
Maureen Batkin. Richard Dennis. The Old Chapel, Shepton Beauchamp, Somerset TA19 0LE
Housewife 1947:
Pocket Publications Ltd, Hutton Press Ltd, 43/44, Shoe Lane, London EC4
Illustrated 19/12/1953:
Odhams Press Ltd. Long Acre, London WC2
John Bull 9/9/1950:
Odhams Press Ltd., Long Acre, London WC2
Meccano Magazine Aug 1950/Feb 1951:
Meccano Ltd., Binns Road, Liverpool 13
Meister Des Puppenspiels (undated):
Deutsches Institut für Puppenspiel, Bochum, Schriftleitung, Fritz Wortelmann
'Puppetually Yours' Pelham Puppet ID & Collectors' Guide
David Leech Productions 1996
Puppets Progress 1953
Jan Bussell, Faber & Faber Ltd., 24 Russell Square, London
Robinson & Cleaver
Christmas gifts Catalogue c1953
Sunday Graphic November 1949
Publisher unknown
Television Weekly Vol. 2 No. 40 1/12/50:
A Precinct Publication, 50, Old Brompton Rd, London SW7
The British Puppet & Model Makers Guild:
Newsletter no.6 Sept 1956

The Daily Telegraph Party Book c1954:
   The Daily Telegraph, 135, Fleet St, London EC4
The Puppet Master June 1956 Vol. 5 no 1:
   Journal of the British Puppet & Model Theatre Guild, Editor J Bussell.
The Sketch 9/11/49:
   The Illustrated London News & Sketch Ltd., Milford Lane, London WC2
The Times 24/4/93:
   News International Newspapers Ltd 1 Virginia Street, London E98 1XY
Through Wooden Eyes 1956:
   Jan Bussell, Faber & Faber Ltd., 24 Russell Square London
TV Comic 1952/60:
   Beaverbrook Newspapers Ltd., Fleet St, London EC4
W. Fieldon & Sons Ltd catalogue 1955/56:
   Percy Bros Ltd., Hotspur Press, Manchester & London
World's Fair Magazine 12/12/1953: World's Fair Ltd., Oldham, Lancashire

WEB SITES:

www.easyweb.easynet.co.uk (MOMI)
www.huntleyandpalmers.org.uk (Huntley & Palmer tins)
www.littlemesters.com (Sheffield steel/ Cutler companies)
www.luckybears.com (Chiltern Hygienic toys)
www.newarkadvertiser.co.uk (Muffin's links with Newark)
www.members.westnet.com.au (Bussell history)
www.myottcollectorsclub.com (Wade Heath information)
www.oxforddiecast.co.uk (TV Comic information for 1961)
www.rushes.co.uk/dizzee_rascal/mainframe.htm (music artist)
www.teddy-bear-uk.com (Chad Valley information)
http://www.whimsicalwades.com (Wade Ceramics)
http://www.windupgram.co.uk (gramophone information)

www.muffin-the-mule.com

To join the club phone Adrienne Hasler on 020 8504 4943

Index

Actifwear, 104
Advertising, 4, 18, 81, 111
Alexandra Palace, 55, 57, 71
Andrews, Julie, 30
Andy Pandy, 92, 104
Annuals, TV Comic, 83, 86
Anti Cyclone, 14, 18
Apron (Coronation), 81, 113
Argosy, 87
Attlee, Mrs Clement, 20
ATV, 15, 16, 24
Australia, 13, 14, 31, 43, 48, 56, 60, 61, 65, 66, 109, 110
Badges (Muffin Club and Road Safety), 81
BAFTA, 66
Bagatelle, 79, 82
Baird, John Logie, 10, 14, 41
Balloons, 82
Bandwagon, 18, 25, 97, 115
Barrett & Sons, 87
Barretts of Staffordshire, 98
Barrie, Amanda, 30
BBC, 4, 5, 7, 9, 10, 14, 15, 20, 23, 24, 25, 30, 31, 33, 35, 37, 45, 55, 57, 69, 83, 103, 115
BBC Children's Hour Annual, 83
Beaverbrook Newspapers, 48, 86, 116
Blake, Molly (Bud), 6, 23, 24, 82, 83, 90, 97
Blue Muffin Book, The, 82, 111
Bob the Builder, 73
Bobby Howes, 69
Books, 68, 82, 111
Bookworms, The, 31
Bracelet charm, 84
British Council, 60
British Puppet and Model Theatre Guild, The, 16, 24
Brooch, 84
Brooch (plastic and enamel), 84
Budleigh Salterton, 29, 32
Bush television, 81
Bussell, Jan, 5, 9, 10, 13, 14, 15, 16, 17, 20, 24, 31, 32, 34, 39, 41, 43, 44, 45, 54, 84, 85, 94, 96, 107, 115, 116
Bussell, Sally, 32
Bussells, 6, 13, 14, 15, 22, 30, 31, 34, 41, 43, 107
Busselton, 13
Busy Box, 84
Buttons (Jason Buttons, London SE24), 84
Calendar, 85
Caleys, 86, 112
Cambridge Theatre, 64
Cards (Birthday, Christmas and Party), 85, 92, 105
Caricature Theatre Company, 72
Ceramic model, 85
Chad Valley, 78, 82, 84, 86, 87, 88, 90, 92, 96, 97, 101, 116
Champion Story Book, 83
Chelsea Palace, 70
Chequers, 59
Children's Press, The, 91
Chiltern Hygienic Toys, 89
Churchill, Sir Winston, 48
Classic Toys Magazine, 89
Collectables, 63, 103
Collins Children's Annual, 83
Colombo (Sri Lanka), 66
Comics and Annuals, 48, 86, 107
Cone Ripman School, 30
Cook, Sue, 71
Coronation, The, 55, 81, 113
Crown (wallpaper), 99
Croydon, 88
Crumpet the Clown (Tickler), 22, 54
Cutlery, 86
Daily Mail, 18, 110
Dalston, 28, 88
Decca (records), 95, 115
Deep Depression, 14, 18
Derby Race Game, 87
Devon Resource Centre, 63
Dr Who night, 33

Drake, Sir Francis, 29
Drawing Slate, 87
Duncan, Sylvia, 25
Edmonds, Noel, 69
Elizabeth II, Queen, 43
Elizabeth, Queen, 43, 81
Embroidery Set, 88
Enid Blyton's Magazine, 96, 99
Ephemera, 88
Eros New Theatre (Piccadilly), 88, 110
Evening Standard, 85
Fabric, 88, 89, 113
Felix, Geoff, 92
Flower Ballet, 14
Flowerpot Men, 92
Freshfields, 48, 51, 56, 71
Friends of Lyndon House, 48, 51
George Rowney & Co, 91
Gilson, Alfred, 11, 27, 28, 93
Glevum, 86
Golden Jubilee, 67
Grace the Giraffe, 36, 95, 98
Gramophone, 89
Green Muffin Book, The, 82
Hair ribbon, 110
Hall Edwin, Ltd, 85
Hamleys, 94
Hammersmith, 19, 88
Hampton Court, 30
Handkerchiefs (Television Novelty Box), 89
Hansard, 68, 103
Harrods, 4, 94
Hasler, Adrienne, 5, 113
Hellzapoppin, 21
Hepburn, Audrey, 19, 79, 88
Here Comes Muffin, 82, 111
Hibbert, Jimmy, 73
Hislop, Ian, 67
Hogarth Puppet Circus, The, 70
Hogarth Puppet Orchestra, The, 70
Hogarth Puppet Theatre, The, 107
Hogarth Puppets, The, 13, 54
Hogarth, Ann, 5, 6, 9, 10, 13, 14, 15, 17, 18, 19, 20, 22, 23, 24, 29, 32, 34, 37, 41, 45, 48, 54, 55, 61, 63, 82, 83, 84, 85, 90, 93, 95, 104, 107, 112

Hogarth Theatre Tent, The 88, 111
Hogarth, William, 34, 85
Hoopoo's Circus, 74
Hubert the Hippo, 36, 90, 96, 98
Huntley, Boorne and Stevens (Huntley and Palmers), 98
Illustrated Magazine, 23
Jackson, Pauline, 9, 31, 64
James Valentine Publishing Company, 85
Jigsaw Puzzle, 90
John Bull, 23
Katy the Kangaroo, 43, 56, 95, 98
Kirri (Kerri) the Kiwi, 13, 24, 85, 98
Kohnstam, Richard, 28, 93
Lamp, 90
Lancaster, Edith, 44
Landau, Cecil, 19, 88
Laye, Evelyn, 69
Lesney, 28, 79, 93
Lime Grove Studios, 20, 32, 57
Little Grey Rabbit, 14, 31
Little White Bull, 94
London, 13, 14, 16, 20, 25, 27, 28, 30, 41, 43, 56, 59, 74, 82, 83, 84, 87, 88, 89, 91, 94, 95, 115, 116
London Records, 95
London Marionette Theatre, 41
London Zoo, 56
Louise the Lamb, 43, 54, 95, 96, 98
Ludo Game, 90
Luntoy (also Luntoy Supertoys), 87, 100, 101
Lyric Theatre (Hammersmith), 19, 88
Macbeth, 14
Maile, Stanley, 9, 43, 59, 82
Main, Neville, 9, 16, 24, 37, 38, 82, 83, 86, 92, 104, 110, 112
Margate, 88
Marks, Alfred, 70
Martin Lucas Ltd, 93
Maverick Entertainment Group Ltd, 37, 39, 49, 50, 107
McNally, Derek, 4, 5, 6, 24, 32, 45, 48, 85, 107, 112
McNally, Sally (see also Bussell, Sally), 5, 10, 15, 18, 19, 29, 47, 53, 54, 104, 106, 107

INDEX

Mills, Annette, 6, 7, 10, 11, 15, 18, 19, 20, 21, 22, 23, 24, 25, 27, 29, 33, 43, 47, 48, 54, 55, 56, 57, 58, 59, 60, 61, 62, 69, 70, 72, 74, 81, 82, 83, 84, 85, 88, 90, 92, 93, 95, 97, 103, 104, 107, 109, 110, 111, 112
Mills, Sir John, 21
Mini-cine projector, 110
Moko Muffin Junior, 4, 11, 26, 27, 28, 79, 93, 107
Moko Peregrine, 93
Moko Toy, 90
Money box, 90
Monty the Monkey, 43, 90, 95, 96, 98
Morris and Doris, 54
Morris, William, 68
Moscow Muffin, 41, 42
Muffin and Louise, 83
Muffin and Peregrine, 83
Muffin and the Reluctant Carrot, 9, 36
Muffin and the Sea Serpent, 86
Muffin Climbs High, 83
Muffin Club, The, 112
Muffin Fishing, 93
Muffin Makes Magic, 83
Muffin On Holiday, 83
Muffin Playing Cricket, 93
Muffin Syndicate, The, 6, 25, 28, 82, 84, 85, 86, 87, 88, 89, 90, 91, 92, 93, 94, 96, 98, 99, 101, 107
Muffin the Detective, 86
Muffin the Mule, 4, 5, 7, 9, 10, 13, 16, 19, 23, 27, 28, 33, 34, 35, 36, 43, 47, 49, 50, 53, 54, 55, 57, 67, 68, 70, 78, 86, 88, 90, 92, 93, 94, 95, 96, 97, 98, 100, 101, 103, 106, 107, 112
Muffin the Mule Collectors' Club, 4, 5, 10, 16, 42, 44, 47, 92
Muffin's ABC book, 83
Muffin's Birthday, 83
Muffin's Caravan Journey, 93
Muffin's Circus, 74
Muffin's Cure for a Cold, 86
Muffin's Own Story Book, 83
Muffin's Picnic, 93
Muffin's Splendid Adventure, 82
Muffin's Thinking Cap, 83
Napkin Ring, 91
National Band Gramophone Company, 89
New Zealand, 13, 14, 43, 56, 65, 109
Newark, 13, 116
Newell and Sorrell, 68
News of the World Ltd, The, 86
Norman, Barry, 67
North Weald, 21, 47
Obratsov, Sergei, 41
Odell, Jack, 28
Ollive, Richard, 73
Opaque Glass, 98
Oriole Records, 95
Oswald the Ostrich, 43, 54, 57, 71, 90, 95, 98
Otto the Octopus, 10, 95
Painting Book, 91
Parthian Films, 15
Patterns (knitting and sewing), 91
Pebble Mill, 19, 56, 71
Pelham Puppet, 31, 84, 115
Pelham Puppets, 31, 84
Pelham, Bob, 94
Pelpups, 31
Pencils, 91
Penknife, 86, 91
Peregrine the Penguin (see also Mr Peregrine Esq), 17, 22, 36, 43, 54, 71, 87, 90, 95, 96, 98
Peter the Pup, 36, 43, 54, 95, 96, 98
Phillips, Jane, 10, 30, 34, 38, 72
Piccadilly, 66, 88, 110
Picture Cubes, 87, 92
Pifco, 86, 90
Playing Cards, 92
Polly the Parrot, 95
Postcards, 92
Pouffes, 93
Prince Charles, 67
Prince, The Happy, 14, 73
Punch and Judy Man, The, 92
Puppet Guild, 18
Puppet Master, The, 13, 16, 116
Puppets and I, The, 96
Puppet's Progress, 9, 84, 94

Purple Muffin Book, The, 82
Queen of Hearts, The, 69
Radio Devon, 62
Radio Newcastle, 67
Radio Times, 45, 66, 67, 69
Rank Studios (Lime Grove), 57
Rascal, Dizzee, 7
Rayner Sisters, 60, 61
Records, 25, 95, 115
Red Muffin Book, The, 41, 82
Reiniger, Lotte, 14, 72
Rhodes, Gary, 67
Ribena, 73
Riverside Theatre, 31
Robbens, Paul, 38, 105
Roberts Bros., 86, 87
Robinson & Cleaver, 89, 115
Rookery, The (Streatham Common), 88, 111
Ross, Jonathan, 67
Royal Mail, 68, 103
Sacul, 87
Sally the Seal, 36, 54, 95
Sam the Scarecrow, 95
Scraps, 95
Selfridges, 4, 28, 100
Sheffield, 88, 110, 116
Sheffield News Theatre, 110
Sibley, Antoinette, 30
Sidmouth, 37
Skittles, 96
Slippers, 96
Smith, Leslie, 28
Smith, Rodney, 28
Snow, Peter, 57
Soaps (figural and tablet), 96
Song Book (First Muffin Song Book), 25, 97
Sooty, 92, 111, 112
South Africa, 14, 20, 21, 25, 30, 31, 43, 65, 109
Southampton, 88
SS Orantes, 110
SS Strathnaver, 60
St Muffin and the Dragon, 63
Stirling, John, 60, 72
Sunday Graphic, 28, 100, 115
Tablecloth, 97
Tankard, 97

Tea set, 97
Television Annual, 83
Television Weekly, 84, 91, 98, 115
Telly Addicts, 69
Tempest, Jack, 63
The Hogarth Theatre Tent, 88, 111
Theatre Royal (Bath), 18
Thomas Turner and Co., 91
Through Wooden Eyes, 31, 41
Tickler's Circus, 74
Tilbury Dockers, 61
Tins (biscuit), 98
Titchmarsh, Alan, 66
Tooting, 88
Transfers, 99, 101
TV Cine projector, 93, 110
TV Comic, 47, 48, 81, 86, 88, 89, 91, 92, 93, 95, 96, 98, 100, 107, 109, 110, 111, 112, 116
TV Mini Books, 86
TV Mirror 21, 22
TV Weekly, 19, 96
UNIMA, 16, 39
Uttley, Alison, 14, 31
Vaudeville Theatre, 88, 95, 110
Wade, Heath & Co (Wadeheath), 97
Wash day, 95
Waterloo News Theatre, 88, 110
Waterlow Park (Highgate), 88, 111
We want Muffin, 54
Weldon, 91
Wend-al, 87
Wheeler, Jimmy, 69
White Barn, The, 32
Whitehead, Jack, 9, 74, 75
Wilde, Oscar, 14, 73
William Morris, 68
Wimbledon Theatre, 69, 88
Wooden toys, 100
Woodentops, 92
Woolacombe Beach, 31, 74
Zebbie the Zebra, 20, 43, 90, 95, 96, 98